WOOLCRAFT

EDITED BY
Liz Bloor

BBC BOOKS

Network books is an imprint of BBC Books,
a division of BBC Enterprises Limited
Woodlands, 80 Wood Lane, London W12 0TT

First published 1993

ISBN 0 563 36794 6

Designed by Roger Daniels

All illustrations by Kate Simunek and all annotations by Mike Gilkes
except Kaffe Fassett charts pp. 95–101 by William Briggs.

Colour photographs by John Quinn
Styling by Michelle Thompson
Photographs taken on location at Merewood Country House Hotel,
Ecclerigg, Windermere, Cumbria
Models supplied by MMA/Manchester Model Agency

Black and white photographs by John Jefford

Set in Goudy by Selwood Systems, Midsomer Norton
Printed and bound in Great Britain by Butler & Tanner Ltd, Frome
Colour separations by Radstock Reproductions
Cover printed by Clays Ltd, St Ives plc

ACKNOWLEDGEMENTS

The Editor and Publishers would like to thank all the contributors for their permission to reproduce copyright material: John Allen; Debbie Bliss; Lesley Conroy; Melinda Coss; Rohana Darlington; Anna Davenport; Ann Davies; Mary Davis; Sally-Anne Elliott; Kaffe Fassett for Autumn Roses Cushion reproduced from *Glorious Inspiration* (£20) by kind permission of Ebury Press; Marion Foale; Glorafilia; Ariella Green; Vikki Haffenden; Gail Harker; Julie Hasler; Ruth Herring; Hilary Highet; Julia Jones; Gary Kennedy for Postman Pat Jumper (Postman Pat © Woodland Animations Ltd, 1993); Paddy Killer; Mary Konior; Linda McDevitt; Freda Parker; Linda Parkhouse; Sophie Schellenberg; J & J Seaton; Alice Starmore; Lois Vickers; Angela Wainwright; and Jane Walmsley. We would also like to thank the staff of the Merewood Country House Hotel, Ecclerigg, Cumbria for their kind and good-humoured help during the photography shoot.

Contents

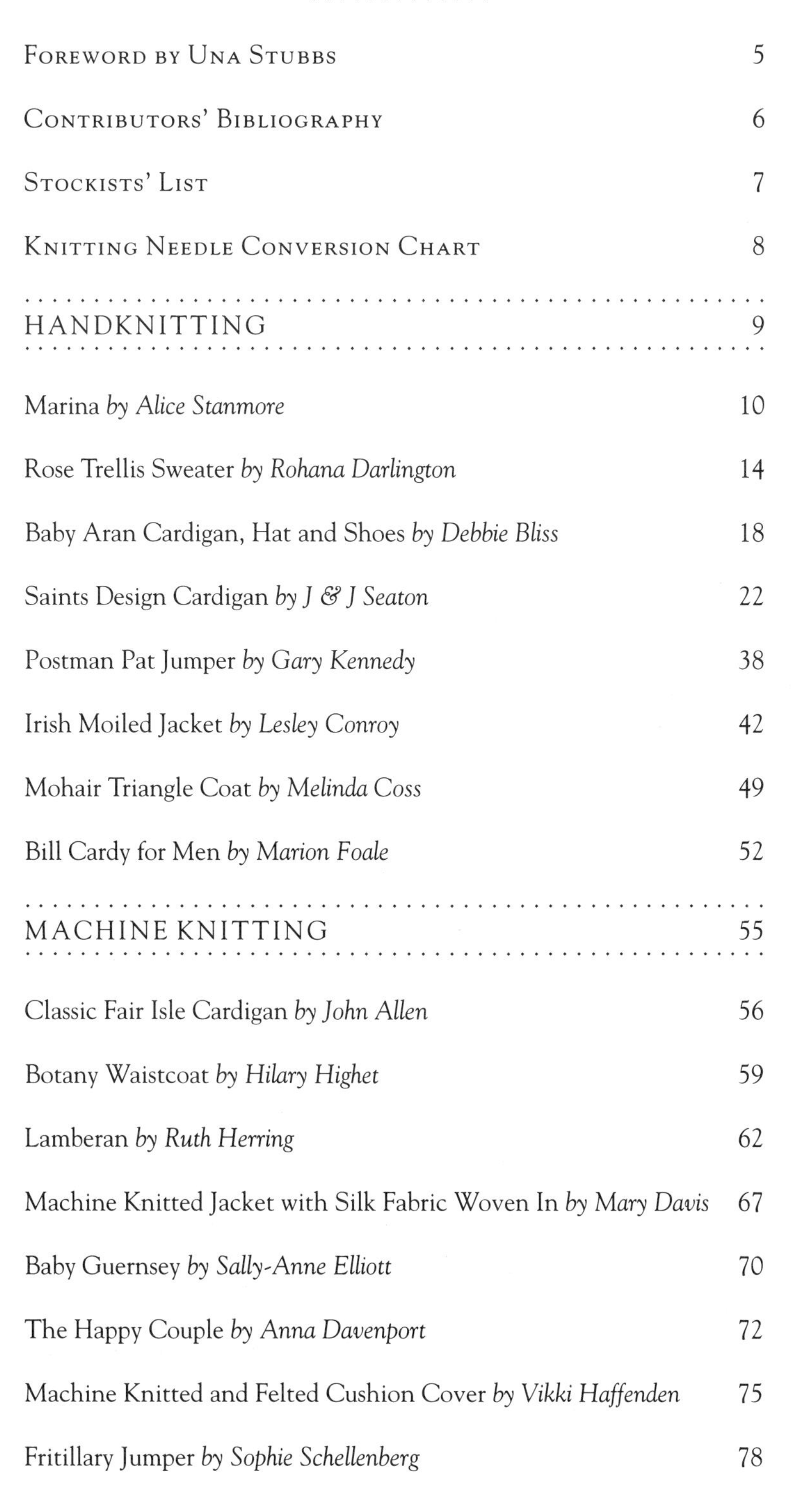

FOREWORD

Crafts have been popular for centuries – not only as a hobby or interest but as a skill for making essential clothing. Today, knitting, crocheting and stitching skills are a means of expressing creativity and provide an opportunity to explore the fascinating world of fabrics, yarns, textures and colours. As well as the traditional hand-craft skills, *Woolcraft* also explores machine-made designs with garments for machine knitting and embroidery.

This book gathers together the work of some of the most popular craft designers practising today, along with their previously unpublished and new designs. Featured are Kaffe Fassett with a needlepoint cushion, J & J Seaton with a stunning hand-knitted cardigan, Marion Foale with an exclusive designer knit and Hilary Highet with a superb machine-knitted waistcoat.

Other subjects covered in this book are crochet, embroidery, cross-stitch, rug-making, patchwork and machine embroidery. With such an array of designs, fabrics and colours there is undoubtedly something here for everybody – and perhaps this inspiring collection will encourage you to experiment with a new craft.

Contributors' Bibliography

JOHN ALLEN
The Machine Knitting Book, Dorling Kindersley, 1985
The Treasury of Machine Knitting, David & Charles, 1989
Fabulous Fair Isle, Lochar, 1991

DEBBIE BLISS
Wild Knitting (contributor), Mitchell Beazley, 1978
She Book of Quick Fashion Clothes, (co-author) Ebury Press, 1984
Good Housekeeping Knitting in Style, (contributor/editor) Ebury Press, 1985
Good Housekeeping Knits for Kids, (contributor/editor) Ebury Press, 1986
Knit One, Style One (contributor), Orbis, 1986
Quick Knits, (co-author) Bell & Hyman, 1986
Baby Knits, Ebury Press, 1988
Country Knits, (co-author) Ebury Press, 1991
Traditional Knitting from the Scottish and Irish Isles, (editor), Ebury Press, 1991
New Baby Knits, Ebury Press, 1991
Kids' Knits for Heads, Hands and Toes, Ebury Press, 1992

MELINDA COSS
Disney Knits, Sidgwick & Jackson, 1987
Knitting with Cotton, Sidgwick & Jackson, 1988
Knitting with Mohair, Sidgwick & Jackson, 1988
Cat Knits, Aurum Press, 1988
Popeye and Friends, Sidgwick & Jackson, 1989
Teddy Knits, Aurum Press, 1989
Magic Carpets, Collins, 1989
Art Deco Knits, Blandford Press, 1990
Folk Knits, Anaya, 1992
Floral Needlepoint, Anaya, 1992
Bloomsbury Needlepoint, Ebury Press, 1992
Big Softies, Aurum Press, 1992
Wild Animal Knits, Cassell, 1993
Floral Cross Stitch, Anaya, 1993
101 Greetings Cards and How To Make Them, Aurum Press (autumn) 1993 (author)

ROHANA DARLINGTON
Irish Knitting, A & C Black, 1991

ANN DAVIES
Rag Rugs, Letts, 1992

ANNA DAVENPORT
Editor of *Needlework* magazine (monthly), Litharne Ltd.
(Previously editor of *Toy Collections*)

MARY DAVIS
Quick Thick Machine Knits, David & Charles, 1990

SALLY-ANNE ELLIOTT
Creative Machine Knitting, Windward, 1988

KAFFE FASSETT
Glorious Knitting, Century, 1985
Glorious Needlepoint, Century, 1987
Glorious Colour, Century, 1988
Family Album, Century, 1989
Glorious Inspiration, Century, 1991

GAIL HARKER
Embroidery Skills: Machine Embroidery, Merehurst, 1991
Fairy Tale Quilts and Embroidery, Merehurst, 1992
The Daisy Book, Garland Press, (June) 1993

JULIE HASLER
Cats and Kittens' Charted Designs, Dover, 1986
Kate Greenaway Alphabet Charted Design, Dover, 1986
Peter Rabbit Iron-on Transfer Patterns, Dover, 1987
Wild Flowers in Cross Stitch, Blandford Press, 1988
Dogs and Puppies in Cross Stitch, Blandford Press, 1988
Kate Greenaway Cross Stitch Designs, David & Charles, 1989
Kate Greenaway Iron-on Transfer Patterns, Dover, 1990
The Little Tale of Benjamin Bunny, Dover, 1990
The Little Tale of Tom Kitten, Dover, 1991
Children's Charted Designs, Dover, 1993
Egyptian Charted Design, Dover, 1993
Needlepoint Designs, Cassell, 1991
The Crafty Cat Workbasket, David & Charles, 1991
Cats and Kittens in Cross Stitch, Blandford Press, 1992
Wild Animals in Cross Stitch, Blandford Press, 1993
Teddybears in Cross Stitch, Merehurst, 1993

RUTH HERRING
Knitting Masterpieces, Pavilion Books, 1987
Knitting Wildlife, Pavilion Books, 1989

JULIA JONES
Cattern Cakes and Lace, Dorling Kindersley, 1987
The National Trust Calendar of Garden Lore, Dorling Kindersley, 1989
Royal Pleasures and Pastimes, David & Charles, 1990
The English Country Craft Collection, David & Charles, 1991
Beads, A & C Black, 1993

GARY KENNEDY
Over 75 booklets featuring famous cartoon characters, available by mail order from Intarsia.

See Stockists' List.

MARY KONIOR
A Pattern Book of Tatting, Dryad Press, 1985

Heritage Crochet – An Analysis, Dryad Press, 1987
Tatting in Lace, Dryad Press, 1988
Tatting Patterns, Batsford, 1989
Crochet Lace, Blandford Press, 1991
Tatting with Visual Patterns, Batsford, 1992

CAROLE LAZARUS
Glorafilia Needlepoint Collection, David & Charles, 1989
Glorafilia Venice Collection, Conran Octopus, 1991

FREDA PARKER
Victorian Embroidery, Anaya, 1990
Victorian Patchwork, Anaya, 1992

ALICE STARMORE
Charts for Colour Knitting, Windfall, 1992

LOIS VICKERS
Embroidery (contributor), Octopus, 1983
The Embroidered Garden, Pyramid, 1989
The Scented Lavendar Book, Ebury Press, 1991
The Daisy Book, Garland Press, (June) 1993

ANGELA WAINWRIGHT
Angela's Flowers, Framecraft Miniatures Ltd, 1989
Jar Lacy's, Framecraft Miniatures Ltd, 1989
Counted Cross Stitch, Letts, 1990
Angela's Small Flowers, Framecraft Miniatures Ltd, 1990
Grapes and Roses, Framecraft Miniatures Ltd, 1990
Anniversary, Framecraft Miniatures Ltd, 1990

JANE WALMSLEY
Patchwork, Letts, 1990
Appliqué, Letts, 1992

Stockists' List

HANDKNITTING

J & J Seaton kits from:
Goetre, Llanfynydd, Carmarthen, Dyfed, SA32 7TT.
Tel: (0558) 668825

Melinda Coss mohair & kits from:
Ty'r Waun Bach, Gwernogle, Camarthen, Dyfed, SA32 7RY.

Gary Kennedy designs and kits available from:
Intarsia, PO Box 138, Uxbridge, Middlesex, UB8 2XR.
Tel: (0734) 595047

Alice Starmore yarns packs from:
The Mission House, Achmore, Isle of Lewis, Western Isles, PA86 9DU.

Luxury Shetland Tweed yarn used by Debbie Bliss. List of stockists from:
Hayfield Textiles Ltd, Hayfield Mills, Glusburn, Keighley, W. Yorkshire, BD20 8QP.

Lesley Conroy kits from:
5 Fagley Road, Undercliffe, Bradford, West Yorkshire, BD2 3LS.

Kits by Ruth Herring, send sae to:
Herring, PO Box 774
London SW11 1EX

MACHINE KNITTING

Colinette Yarn (Mary Davis) from:
Colinette Yarns, Park Lane House, High Street, Welshpool, Powys, Wales.

Ruth Herring recommends:
The Yorkshire Mohair Mill, Unit 1, Gibson Street, Bradford, BD3 9TS.

Vicki Haffenden recommends:
Jamieson and Smith, Shetland Wool Brokers, 90 North Road, Lerwick, Shetland, ZE1 0PQ.

Pamela Wise Designer Yarns, Tanfield Lane, Wickham, Fareham, Hants, PO17 5NS.

Rowan Yarns Limited, Green Lane, Holmfirth, West Yorkshire.

Texere Yarns, College Mill, Barkerend Road, Bradford, West Yorkshire, BD3 9AQ.

Brockwell Wools, Stanfield Mill, Triangle, Sowerby Bridge, HX6 3LX.

Anna Davenport recommends Bramwell Yarns. For your nearest stockist contact:
FW Bramwell & Co. Ltd, Unit 5, Metcalf Drive, Altham Lane, Altham, Accrington, BB5 5TU
Tel: (0282) 779811.

Chenille Yarns used by John Allen and also recommended by Hilary Highet:
Yeomans Yarns Ltd, 36 Churchill Way, Fleckney, Leicester, LE8 0UD.

CROCHET

Mary Konior's yarns from:
Coats Leisure Crafts Group, 39 Durham Street, Glasgow, G41 1BS.

Steel hooks from:
Crochet Design Centre, 17 Poulton Street, Morecambe, Lancs, LA4 5PZ and Hollyoak Crochet, Cogsall Lane, Comberbach, Cheshire, CW9 6BS.

STITCHING

Kaffe Fassett kits can be obtained from:
Ehrman, 21/22 Vicarage Gate, London W8 4AA.

Glorafilia kits can be obtained from:
Glorafilia, The Old Mill House, The Ridgeway, Mill Hill Village, London NW7 4ED.

Angela Wainwright's Christening sampler bookmark, pot and spoon available from:
Framecraft Miniatures Limited, 148–150 High Street, Aston, Birmingham.

Angela Wainwright Designs in kit form available, with catalogue, from:
8 Danes Court Road, Tettenhall, Wolverhampton, WV6 9BG

Lois Vickers recommends DMC Embroidery threads by mail order from:
Campden Needlecraft Centre, High Street, Chipping Campden, Gloucestershire, GL55 6AG.

RAG RUGS

Materials obtainable from:
Ann Davies, 1 Wingrad House, Jubilee Street, London E1 3BJ

Knitting Needle Conversion Chart

Metric	*English*	*American*
2mm	14	0
$2\frac{1}{4}$mm	13	1
$2\frac{3}{4}$mm	12	2
3mm	11	–
$3\frac{1}{4}$mm	10	3
$3\frac{3}{4}$mm	9	5
4mm	8	6
$4\frac{1}{2}$mm	7	7
5mm	6	8

Metric	*English*	*American*
$5\frac{1}{2}$mm	5	9
6mm	4	10
$6\frac{1}{2}$mm	3	$10\frac{1}{2}$
7mm	2	–
$7\frac{1}{2}$mm	1	–
8mm	0	11
9mm	00	13
10mm	000	15

HANDKNITTING

The British Isles are renowned for their strong tradition of handknitting, from the highly intricate stockings knitted in wools and silks of the sixteenth century to the traditional fisherman's ganseys and Fair Isles of the nineteenth century. These essential crafts have now developed into an art form. Our designers produce top-quality garments in beautiful yarns which incorporate some of the oldest traditions with innovative design. The handknitting section of the book features some of Britain's best knitwear designers and shows a vast array of techniques and designs.

Using traditional techniques are Alice Starmore with her beautiful Fair Isle, Rohana Darlington with her Aran Sweater and Debbie Bliss, who brings a new twist to Aran using unusual colours in her design for babies. Lesley Conroy concentrates on the original wools with her stunning animal knit and Marion Foale brings a high fashion element with her simple gansey.

Jamie Seaton's Saint's cardigan uses intarsia to stunning effect whilst Melinda Coss takes a fresh look at mohair. And Gary Kennedy's simple but very effective Postman Pat design is one that all children will love.

Marina

BY ALICE STARMORE

A complicated pattern which is not for the inexperienced or faint hearted, this is traditional Fair Isle at its best, knitted in the round on small needles with lots of colour stranding and cut open to form the shape.

MEASUREMENTS

Sizes – to fit bust 86–91cm(94–99cm)/34–36in(37–39in)
Directions for second size are given in brackets. Where there is only one set of figures, it applies to both sizes.
Underarm (buttoned) 104(114)cm/41(45)in.
Length from top of shoulder 53(57)cm/21 ($22\frac{1}{2}$)in.
Sleeve length 48(49)cm/19($19\frac{1}{2}$)in.

MATERIALS

Shetland jumper weight 1oz skeins as follows:

Deep Aqua	2(3)
Crowberry, Salmon	2
Sage, Burnt Heather, Indigo, Coral, Blue Grass, Dark Green	1(2)
Thistle, Green Mist, Wood Green, Dog Rose, Evergreen, Corn, Blue Mist, Mauve Mist, Fog, Mushroom, Turquoise, Claret, Bright Turquoise	1

One set each of double-pointed or circular 3mm (11) and $3\frac{1}{4}$mm (10) needles
2 small safety pins
1 stitch holder
Stitch markers
1 darner
6 buttons

STITCHES

2/2 Corrugated Rib K2 with first colour, p2 with second colour, stranding yarns evenly across wrong side.

Fair Isle Chart Pattern K every round, and on two-colour rounds, strand the yarn not in immediate use evenly across wrong side. On stretches of more than 8 sts in one colour, weave in yarn not in use at centre of stretch.
Steeks These are worked at front, armholes and neck, and later cut up centre to form openings. To make a steek, cast on 8 sts. K these sts on every round, and on two-colour rounds, k each st and round in alternating colours. Do not weave in newly joined in or broken off yarns at centre of front steek. Instead, leave approx 5cm (2in) tail when joining in and breaking off yarns.
Edge stitch Worked at each side of front, neck and armhole steeks and k in background colours on all rounds. Sts for sleeves and front and neck band are knitted up from edge sts.
Cross-stitch With darner, overcast raw edge of steek to strands on wrong side, and after sewing to end, reverse to form cross-stitches.

TENSION

16 sts and 16 rows for a 5cm (2in) square, measured over Fair Isle chart pattern, using $3\frac{1}{4}$mm (10) needles. To make an accurate gauge swatch, cast on 33 sts on 1 double-pointed or circular needle and work on a flat piece, *knitting on the right side only*, breaking off the yarns at the end of every row.

ABBREVIATIONS

alt–alternate, **beg**–begin(ning), **cont**–continue, **dec**–decrease, **foll**–follow(ing), **k**–knit, **patt**–pattern, **p**–purl, **rem**–remaining, **rep**–repeat, **rnd**–round, **ssk**–slip, slip and knit the two slipped stitches together, **st(s)**–stitch(es), **tog**–together.

Body

With $3\frac{1}{4}$mm (10) needles and Deep Aqua, cast on 328(360) sts. Place a marker at beg on rnd, join in Claret and work front steek, edge sts, and 2/2 corrugated rib as folls:

Rnd 1 With alt colours, k4 steek sts; with Deep Aqua, k1 edge st, *k2 Deep Aqua, p2 Claret; rep from * to the last 7 sts; k2 Deep Aqua; with Deep Aqua, k1 edge st; with alt colours k4 steek sts.

Rnds 2 and 3 As set, substituting Coral for Claret.

Rnds 4, 5 and 6 As set, substituting Salmon for Coral.

Rnd 7 As set, substituting Mushroom for Salmon.

Rnd 8 As set, substituting Indigo for Deep Aqua.

Rnd 9 As set, substituting Fog for Mushroom.

Rnd 10 As set, substituting Crowberry for Indigo.

Rnd 11 As set, substituting Mauve Mist for Fog.

Rnd 12 As set, substituting Dark Green for Crowberry.

Rnds 13, 14 and 15 As set, substituting Blue Mist for Mauve Mist.

With Dark Green, k1 rnd and inc 3 sts evenly spaced above corrugated rib. 331(363) sts. Mark first st of rnd (centre of steek) and joining in and breaking of colours as required, begin at rnd 1 and set the patt as follows:

K4 steek sts, k1 edge st in darker colour, rep the 32 patt sts from chart (see p. 13) 10(11) times, patt the last st from chart, k1 edge st in darker colour, k4 steek sts. Continue as set and on next rnd, place markers at underarms as folls:

K4 steek sts, k1 edge st, patt 80(88) sts, patt the next st and mark it, patt 159(175) sts, patt the next st and mark it, patt 80(88) sts, k1 edge st, k4 steeks sts. Continue as set and work all 68 rnds of chart, then work rnds 1 to 3(9) again.

Next rnd – begin V-neck shaping:

Keeping continuity, k4 steeks sts, k1 edge st, ssk, patt to the last 7 sts, k2 tog, k1 edge st, k4 steek sts. Place a marker on each edge st to mark beg of shaping. Cont in patt and dec as set at neck edges on every foll third rnd. *At the same time*, on rnd 10(18) of chart, begin armholes as folls:

Cont as set to first marked underarm st, patt this st and place it on a safety pin. Cast on 10 sts (with alt colours on 2nd size): keeping continuity, patt 159(175) sts of back, patt the next marked underarm st and place it on a safety pin, cast on 10 sts (with alt colours on 2nd size), work as set to end of rnd. Cont as set and work the first and last of each 10 sts cast on in darker colours for armhole edge sts, and rem 8 sts in alt colours for armhole steeks. Continue until 144(156) chart patt rnds have been worked from beg, ending with rnd 8(20) inclusive.

Next rnd – begin back neck shaping:

Keeping continuity, work as set to back chart patt sts, patt 56(62) sts of back, place the next 47(51) sts on a holder; with alt colours, cast on 10 sts (the first and last are edge sts, the centre 8 are steek sts); keeping continuity, patt as set to end of rnd. Work back neck steek in alt colours and edge sts in darker colours and dec 1 st at each side of back neck edge sts on next and every foll alt rnd and continue to dec at front neck on every 3rd rnd until 53(59) chart patt sts rem on each shoulder. Continue straight as set to rnd 16(28) of 3rd chart rnd rep. Patt rnd 16(28), casting off all steek sts on this rnd.

With Evergreen (Crowberry), graft back and front shoulder sts together.

Sleeves

Cut open armhole steeks up centre, between 4th and 5th steek sts. With $3\frac{1}{4}$mm (10) needles and Burnt Heather (Evergreen) pick up and k underarm st from safety pin and place a marker on this st; knit up 144(152) sts evenly around armhole, working into edge sts. 145(153) sts in total. Beg at rnd 24(18) of chart and patt sleeve as folls:

K the marked st (in Evergreen for 2nd size), patt the last 8(12) sts of 32 st rep, rep the 32 patt sts 4 times, patt the first 8(12) sts of chart. Patt next 2 rnds as set, working marked st in darker colours. Shape sleeve as folls:

K the marked st in darker colours. Keeping continuity, k2 tog, patt as set to the last 2 sts, ssk. Keeping continuity, patt 3 rnds straight. Rep these last 4 rnds until 77(83) sts rem.

Second size only Patt 2 rnds straight.

Next rnd Dec for cuff:

With Burnt Heather:

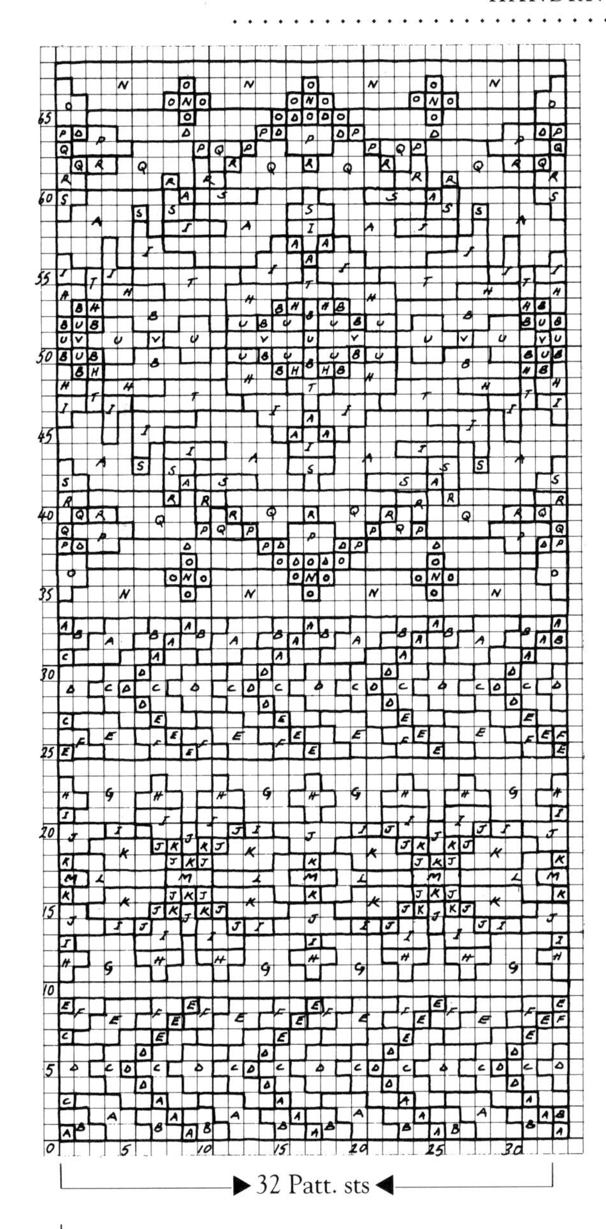

▶ 32 Patt. sts ◀

▶ Last st.

A Deep Aqua
B Blue Grass
C Sage
D Crowberry
E Thistle
F Green Mist
G Burnt Heather
H Coral
I Salmon
J Wood Green
K Dog Rose
L Evergreen
M Corn
N Dark Green
O Blue Mist
P Mauve Mist
Q Indigo
R Fog
S Mushroom
T Turquoise
U Claret
V Bright Turquoise

First size K3, k2 tog; *k4, k2 tog; rep from * to end of rnd. 64 sts.
Second size (K4, k2 tog) 4 times; (k3, k2 tog) 7 times; (k4, k2 tog) 4 times. 68 sts.
Both sizes With 3mm (11) needles work 2/2 corrugated rib in colour sequence as rnd 15 back through rnd 1 of body. With Deep Aqua, cast off knitwise.

Front and Neckband
Cut front and back neck band steeks open up centre, between 4th and 5th steek sts. With 3mm (11) needles and Crowberry, work into edge sts, beg at cast on edge of right front and knit up 72(76) sts to marker at beg of neck shaping; knit up 71(77) sts to back neck holder; pick up and k the 47(51) sts from holder decreasing 1 st at centre back neck; knit up 71(77) sts to marker at beg of left neck shaping; knit up 72(76) sts to cast on edge. 334(358) sts. Work 2/2 corrugated rib in colour sequence as rnd 10 back through rnd 1 of body, beginning and ending wrong side rows with p2, and right side rows with k2. On 6th row work buttonholes as folls:
With right side facing, k2, p2, *cast off 2, k11(12); rep from * 5 times in all; cast off 2; rib to end of row. On next row, cast on 2 sts over those cast off.

Finishing
Trim all steeks to a two stitch width, and with Deep Aqua, cross-stitch steeks to wrong side, covering raw edges. Darn in all loose ends. Sew on buttons.

Rose Trellis Sweater

BY ROHANA DARLINGTON

A difficult Aran for the competent knitter, this superb sweater is well worth the time and effort involved. A variety of techniques and stitches are used to challenge the most accomplished of knitters.

MEASUREMENTS

One size only to fit up to 96cm (38in) bust.
Actual width across back at underarm 55cm ($21\frac{1}{2}$in)
Length from centre back neck to hip 56cm (22in)
Sleeve seam including armhole inset: 46cm (18in)

MATERIALS

Rowan Sea Breeze Soft Cotton in the following colours and quantities

A	6 × 50g	Bleached White	521
B	1 × 50g	Pine Forest	538
C	1 × 50g	Baize	540
D	1 × 50g	Strawberry Ice	546
E	1 × 50g	Antique Pink	533
F	1 × 50g	Sugar Pink	545
G	1 × 50g	Sienna	535

1 pair $2\frac{3}{4}$mm (12) 79cm (31in) circular needles
1 pair $3\frac{1}{4}$mm (10) 36cm (14in) straight needles
1 cable needle
Spare stitch holder

NOTE ON MATERIAL

Colours B–G are used in small quantities, and, as the yarn is only available in 50g balls, there will be quite a bit left over. Some of this can be used in other garments or can be kept to add to your store of yarns for future knitting projects.

TENSION

28 sts and 32 rows for a 10cm (4in) square of Chart B pattern (see p. 17) on $2\frac{3}{4}$mm (12) needles, or size needed to obtain this tension.
34 sts and 34 rows for a 10cm (4in) square of Chart A pattern (see p. 17) on $3\frac{1}{4}$mm (10) needles, or size needed to obtain this tension.

NOTES ON TECHNIQUES

This sweater uses a variety of techniques, sometimes combining them in one part of the garment. The techniques used are intarsia, cabling, ribbing and the trellis pattern worked in garter stitch and purl.
Intarsia Use this technique for the borders, following Chart A on p. 17, and combine with cabling as indicated. Use only short lengths of yarn to avoid tangling, with separate lengths of yarn for each colour as shown in the chart. Twist yarns together at each join to avoid holes, then leave in position to be worked in next row. Knit the ends in as you work by wrapping them round the new colour of thread for two or three sts beyond where they were last used, to minimize later darning in.
Cabling Use cable needle and refer to Chart B instructions on p. 17.

ABBREVIATIONS

alt–alternate, **beg**–beginning, **cont**–continue, **dec**–decrease, **foll**–follow(ing), **inc**–increase, **k**–knit, **patt**–pattern, **p**–purl, **rem**–remaining, **rep**–repeat, **st(s)**–stitch(es), **st st**–stocking stitch.

Back

Beg with Chart A border patt.
Using $3\frac{1}{4}$mm (10) needles and yarn A, cast on 34 sts. Work from Chart A, using intarsia for motifs and cabling for diamond shapes. Work 3 reps of the 52 row patt then cast off. Darn in any loose ends, then wash and block before continuing.

Turn border strip, and using $2\frac{3}{4}$mm (12) circular needles as straight, and yarn A, right side facing, pick up evenly and k 156 sts from along edge (a). Break off yarn and turn so right side faces again. Re-join yarn A, then working from Chart B, rep the 26 sts patt 6 times across the 156 sts. Work the 21 row patt rep 3 times, then work 13 rows of 4th patt rep.
Armhole shaping Cast off 8 sts at beg of next 2 rows (rows 77 and 78). (140 sts).*
At the same time cont in patt until a total of 7 patt reps (147 rows) have been completed. Cast off.

Front

Work as for back until * is reached. Cont in patt until 6 patt reps have been completed (row 126 of Chart B patt).
Divide for neck Right side facing, work across 65 sts of first row of 7th patt rep (row 127). Cast off 10 sts for front neck, then work across 65 sts, cont in patt as set. Transfer first 65 sts to stitch holder, then shape left shoulder as folls:
Row 128 Work with no dec, then cast off 6 sts on each of foll 2 alt (every other) rows (53 sts).
Row 132 Work in patt with no dec.
Row 133 Cast off 4 sts at neck edge (49 sts).
Row 134 Work in patt with no dec.
Row 135 Cast off 2 sts at neck edge (47 sts).
Rep **rows 134 and 135** 3 more times (41 sts).
Row 142 Work in patt with no dec.
Row 143 Cast off 1 st at neck edge (40 sts).
Rep **rows 142 and 143** once more. Work 2 more rows in patt with no further dec, then cast off (39 sts). Transfer rem. 65 sts on stitch holder to $3\frac{1}{4}$mm (10) needles and shape right shoulder similarly, reversing shapings.

Sleeves

Using $3\frac{1}{4}$mm (10) needles and yarn A, cast on 34 sts. Work 2 patt reps of the 52 row border patt from Chart A. Cast off, then darn in any loose ends and wash and block before continuing.
Turn border strip, and using $2\frac{3}{4}$mm (12) circular needles as straight, and yarn A, right side facing, pick up evenly and k 104 sts from along edge (a) for right sleeve and edge (b) for left sleeve. Break off yarn and turn so right side faces again, re-join yarn and then working from Chart B, work 4 patt reps of the 26 st patt across the 104 sts. Work in patt as set, *at the same time* inc 1 st at beg and end of every 8th row, incorporating new sts into extra patt rep at either edge, until 124 sts exist and four 21 row patt reps have been completed. Cast off.

Cuffs

Using $2\frac{3}{4}$mm (12) circular needles as straight and yarn A, right side facing, pick up evenly and k 58 sts from along edge (b) for right sleeve and edge (a) for left sleeve. Set the next row in k1, p1 single rib then work in rib until cuff measures 11cm ($4\frac{1}{2}$in). Change to yarn B and rib 2 rows, then change to yarn F, rib one row, then cast off ribwise in yarn F.

Neckband

Using backstitch, join shoulder seams of back to front.
Using $3\frac{1}{4}$mm (10) needles and yarn A, cast on 4 sts, then work as follows:
Row 1 P1, k2, p1.
Row 2 K1, p2, k1.
Cont in this way making a strip approx 60cm ($23\frac{1}{2}$in) long, to fit snugly into neckline of sweater, then cast off. Beg at left-hand shoulder seam, using invisible seaming and yarn A, sew strip into neckline to form base for later neckband ribbing, using the reverse st st edge of strip to attach to the neck edge of sweater. Join strip together at shoulder seam using backstitch.
Using $2\frac{3}{4}$mm (12) circular needles and yarn A, right side facing, beg at right-hand shoulder seam, pick up evenly and k 70 sts from front neckline and 50 sts from back neck (120 sts) working into the reverse st st edge of band. Work in rounds, and set in k1, p1 rib for 2cm ($\frac{3}{4}$in). Change to yarn B, rib two rounds, then change to yarn F, rib one row, then cast off ribwise in yarn F.

Making Up and Finishing

Darn in any loose ends, then block pattern pieces. Using backstitch, set in sleeves, carefully aligning centre point to shoulder seam. Using backstitch for pattern areas and invisible seam for ribbing, join side and sleeve seams. Press seams lightly on wrong side with warm iron over a damp cloth, then press right side similarly, omitting ribbing.

Colour working

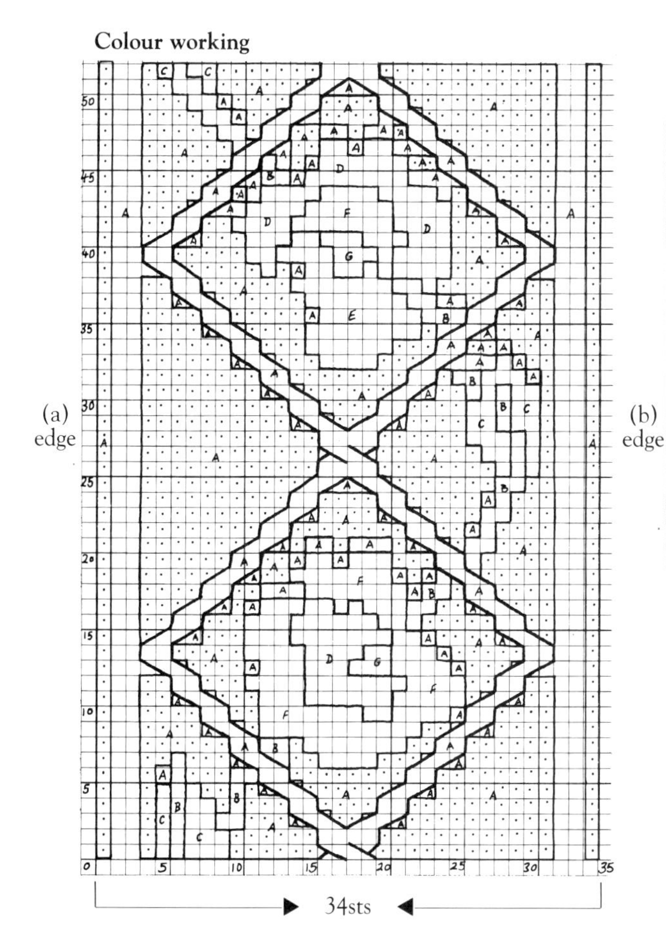

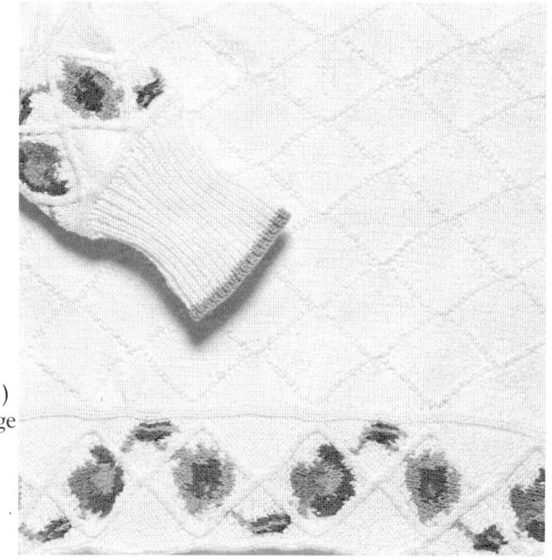

(b)
edge

Cabling Instructions

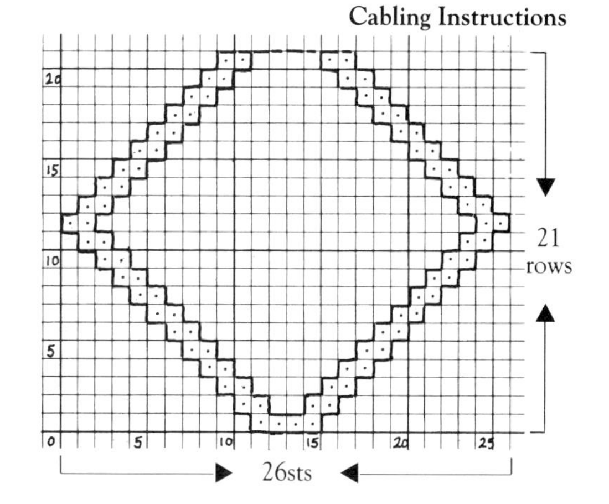

CHART A
· P on RS, K on WS
☐ K on RS, P on WS

A Bleached White
B Pine Forest
C Baize
D Strawberry Ice
E Antique Pink
F Sugar Pink
G Sienna

CHART B
· P on RS
K on WS
☐ K on RS
P on WS

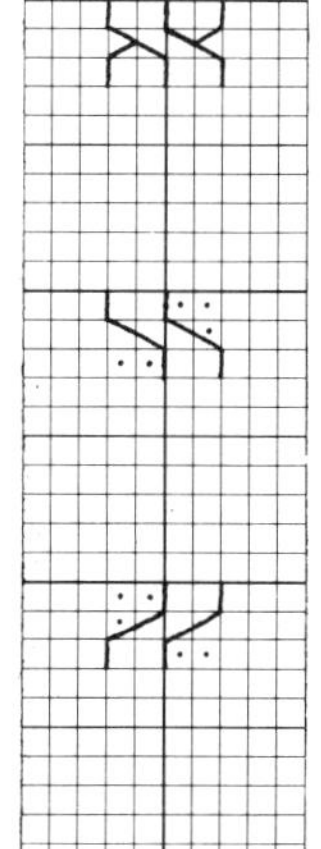

C4F–(cable 2 to front) Slip 2 sts onto cable needle, hold at front, K2, then K2 from cable needle.

C4FP–(cable 2 to front and purl) Slip 2 sts onto cable needle, hold at front, p2, then K2.

C4BP–(cable 2 to back and purl) Slip 2 sts onto cable needle, hold at back, K2 then p2 from cable needle.

Baby Aran Cardigan, Hat and Shoes

BY DEBBIE BLISS

A delightful, fun outfit for babies, this beautiful Aran cardigan, hat and shoes does need some experience to knit, although it may be fun to perfect the craft of Aran knitting with such a tiny garment.

MEASUREMENTS

To fit age 0–3(3–6, 6–9) months
Actual chest measurement 58(63,68)cm/ 23 (25,27)in
Length to shoulder 24(26,29)cm/$9\frac{1}{2}$($10\frac{1}{4}$,$11\frac{1}{2}$)in
Sleeve seam 15(16,18)cm/6($6\frac{1}{4}$,7)in

MATERIALS

Cardigan: 4(4,5) × 50g balls of Hayfield Luxury DK Shetland Tweed in Glen (002)
Pair each of $3\frac{1}{4}$mm (10) and 4mm (8) needles, cable needle, 3 buttons
Hat: 1(1,2) × 50g balls of Hayfield Luxury DK Shetland Tweed in Glen (002)
Pair of $3\frac{3}{4}$mm (9) needles
Shoes: 1 × 50g ball of Hayfield Luxury DK Shetland Tweed in Glen (002)
Pair of $3\frac{1}{4}$mm (10) needles, cable needle

TENSION

22 sts and 28 rows for a 10cm/4in square in st st on 4mm (8) needles.

ABBREVIATIONS

alt–alternate, **beg**–begin(ning), **cont**–continue **dec**–decreas(e)ing **foll**–follow(ing), **inc**–increas(e)ing, **k**–knit, **patt**–pattern, **p**–purl, **rem**–remain(ing), **rep**–repeat, **st(s)**–stitch(es), **st st**–stocking stitch, **tbl**–through back loops, **tog**–together
C3R–slip next st onto cable needle and leave at back, k2, then p1 from cable needle
C3L–slip next 2 sts onto cable needle and leave at front, p1, then k2 from cable needle
C4B–slip next 2 sts onto cable needle and leave at back, k2, then k2 from cable needle
C4F–slip next 2 sts onto cable needle and leave at front, k2, then k2 from cable needle
MB–(Make bobble) pick up loop lying between st just worked and next st and work k1, p1, k1, p1, k1 into it, turn, p5, turn, k5, turn, p2tog, p1, p2 tog, turn, slip 1, k2 tog, pass slipped stitch over.

PANEL A – worked over 20 sts.
1st row (Right side) (p2, C3R) twice, (C3L, p2) twice.
2nd row (K2, p2, k3, p2) twice, k2.
3rd row P1, (C3R, p2) twice, C3L, p2, C3L, p1.
4th row K1, p2, k3, p2, k4, p2, k3, p2, k1.
5th row (C3R, p2) twice, MB, p1 then pass bobble st over this st. P1, C3L, p2, C3L.
6th row (P2, k3) twice, (k3, p2) twice.
7th row (C3L, p2) twice, (p2, C3R) twice.
8th row As 4th row.
9th row P1, (C3L, p2) twice, C3R, p2, C3R, p1.
10th row As 2nd row.
11th row (P2, C3L) twice, (C3R, p2) twice.
12th row K3, p2, k3, p4, k3, p2, k3.
These 12 rows form the pattern.

CARDIGAN

Back

With $3\frac{1}{4}$mm (10) needles cast on 63(69,77) sts.
1st row (Right side) k1, (p1, k1) to end.
2nd row P1, (k1, p1) to end.
Rep last 2 rows 4 times more then work 1st row again.
Inc row Rib 2(1,3), (inc in next st, rib 1(2,3) 1(2,2) times, * rib 1, inc in each of next 2 sts, rib 4 (inc in next st, rib 4) 3 times, inc in each of next 2 sts, rib 1*; (rib 1, inc in next st) twice, rib

1, rep from * to *, (rib 1(2,3), inc in next st) 1(2,2) times, rib 2(1,3). 81(89,97) sts.
Change to 4mm (8) needles. Work in patt as follows:
1st row (Right side) k1, (p1, k1) 2(4,6) times, *p1, k4, p1, work 1st row of panel A, p1, k4, p1*, k1 (p1, k1) 3 times, rep from * to *, k1 (p1, k1) 2(4,6) times.
2nd row P1, (k1, p1) 2(4,6) times, *k1, p4, k1, work 2nd row of panel A, k1, p4, k1*; p1, (k1, p1) 3 times, rep from * to *, p1, (k1, p1) 2(4,6) times.
3rd row P1, (k1, p1) 2(4,6) times, *p1, C4B, p1, work 3rd row of panel A, p1, C4F, p1*; p1, (k1, p1) 3 times, rep from * to *, p1, (k1, p1) 2(4,6) times.
4th row: K1, (p1, k1) 2(4,6) times, *k1, p4, k1, work 4th row of panel A, k1, p4, k1* k1 (p1, k1) 3 times, rep from * to *, k1 (p1, k1) 2(4,6) times.
5th to 12th rows Rep 1st to 4th rows twice but working 5th to 12th rows of panel A.
These 12 rows form patt. Cont in patt until back measures 24(26,29)cm/$9\frac{1}{2}$($10\frac{1}{4}$,$11\frac{1}{2}$)in from beg, ending with wrong side row.
Shape shoulders Cast off 13(15,16) sts at beg of next 2 rows and 13(14,16) sts at beg of foll 2 rows. Leave rem 29(31,33) sts on a holder.

Left Front
With $3\frac{1}{4}$mm (10) needles cast on 31(33,37) sts. Work 11 rows in rib as given for back welt.
Inc row Rib 2, inc in each of next 2 sts, rib 4, (inc in next st, rib 4) 3 times, inc in each of next 2 sts, rib 1 (1,3), (rib 2, inc in next st) 0(2,2) times, rib 5(1,3). 38(42,46) sts.
Change to 4mm (8) needles. Work in patt as follows:
1st row (Right side) (k1, p1) 3(5,7) times, k4, p1, work 1st row of panel A, p1, k4, p2.
2nd row K2, p4, k1, work 2nd row of panel A, k1, p4, (k1, p1) 3(5,7) times.
3rd row (P1, k1) 2(4,6) times, p2, C4B, p1, work 3rd row of panel A, p1, C4F, p2.
4th row K2, p4, k1, work 4th row of panel A, k1, p4, k2, (p1, k1) 2(4,6) times.
5th to 12th rows Rep 1st to 4th rows twice but working 5th to 12th rows of panel A.
These 12 rows form patt. Cont in patt until front measures 14(15,17)cm/$5\frac{1}{2}$(6,$6\frac{1}{2}$)in from beg, ending with wrong side row.

Shape front Keeping patt correct, dec one st at end of next row and 9(8,9) foll alt rows then on every foll 3rd row until 26(29,32) sts rem. Cont straight until front matches back to shoulder shaping, ending with wrong side row.
Shape shoulder Cast off 13(15,16) sts at beg of next row. Work 1 row. Cast off rem 13(14,16) sts.

Right Front
With $3\frac{1}{4}$mm (10) needles cast on 31(33,37) sts. Work 11 rows in rib as given for back welt.
Inc row Rib 5(1,3), (inc in next st, rib 2) 0(2,2) times, rib 1(1,3), inc in each of next two sts, rib 4, (inc in next st, rib 4) 3 times, inc in each of next 2 sts, rib 2. 38(42,46) sts.
Change to 4mm (8) needles. Work in patt as follows:
1st row (Right side) p2, k4, p1, work 1st row of panel A, p1, k4, (p1, k1) 3(5,7) times.
2nd row (P1, k1) 3(5,7) times, p4, k1, work 2nd row of panel A, k1, p4, k2.
3rd row P2, C4B, p1, work 3rd row of panel A, p1, C4F, p2, (k1, p1) 2(4,6) times.
4th row (K1, p1) 2(4,6) times, k2, p4, k1, work 4th row of panel A, k1, p4, k2.
5th to 12th rows Rep 1st to 4th rows twice but working 5th to 12th rows of panel A.
These 12 rows form patt. Complete as given for left front, reversing shapings.

Sleeves
With $3\frac{1}{4}$mm (10) needles cast on 31(33,35) sts. Work 11 rows in rib as given for back welt.
Inc row Rib 2(1,0), (inc in next st, rib 1) 1(2,3) times, rib 1, inc in each of next 2 sts, rib 2, (inc in next st, rib 2) 5 times, inc in each of next 2 sts, rib 1, (rib 1, inc in next st) 1(2,3) times, rib 2(1,0). 42(46,50) sts.
Change to 4mm (8) needles. Work in patt as follows:
1st row (Right side) (k1, p1) 3(4,5) times, k4, p1, work 1st row of panel A, p1, k4, (p1, k1) 3(4,5) times.
2nd row (P1, k1) 3(4,5) times, p4, k1, work 2nd row of panel A, k1, p4, (k1, p1) 3(4,5) times.
3rd row (P1, k1) 2(3,4) times, p2, C4B, p1, work 3rd row of panel A, p1, C4F, p2, (k1, p1) 2(3,4) times.

4th row (K1, p1) 2(3,4) times, k2, p4, k1, work 4th row of panel A, k1, p4, k2, (p1, k1) 2(3,4) times.
These 4 rows *set* the pattern. Cont in patt as set, working appropriate rows of panel, *at the same time*, inc one st at each end of next and every foll 5th row until there are 52(56,62) sts, working inc sts into double moss st (side edge) patt. Cont straight until sleeve measures 15(16,18)cm/ 6($6\frac{1}{4}$,7)in from beg, ending with wrong side row. Cast off.

Front Band and Collar
Join shoulder seams. With $3\frac{1}{4}$mm (10) needles and right side facing, pick up and k31(33,37) sts up straight edge of right front, 37(40,43) sts along shaped edge to shoulder, k29(31,33) centre back neck sts, pick up and k37(40,43) sts down shaped edge of left front and 31(33,37) sts along straight edge. 165(177,193) sts.
Work 3 rows in rib as given for back welt.
1st buttonhole row Rib 3, (cast off 2, rib 9(10,12) sts more) 3 times, rib to end.
2nd buttonhole row Rib to end, casting on 2 sts over those cast off in previous row.
Rib 1 row.
Next row Rib to last 31(33,37) sts, turn.
Next row Slip 1, rib to last 31(33,37) sts, turn.
Next 2 rows Slip 1, rib to last 33(35,39) sts, turn.

Next 2 rows Slip 1, rib to last 35(37,41) sts, turn.
Next 2 rows Slip 1, rib to last 37(39,43) sts, turn.
Next row Slip 1, rib to end.
Rib 2 rows across all sts. Cast off in rib.

To Make Up
Sew on sleeves, placing centre of sleeves to shoulder seams. Join side and sleeve seams. Sew on buttons.

HAT
With $3\frac{3}{4}$mm (9) needles cast on 86(94,102) sts.
1st row K2, (p2, k2) to end.
2nd row P2, (k2, p2) to end.
Rep these 2 rows until work measures 17(19,21)cm $6\frac{1}{2}$ ($7\frac{1}{2}$, $8\frac{1}{4}$) in from beg, ending with 2nd row.
Shape Top
Next row K2, (p2 tog, k2) to end.
Next row P2, (k1, p2) to end.
Next row K2, (p1, k2) to end.
Next row P2, (k1, p2) to end.
Next row K2 tog, (p1, k2 tog) to end.
Next row P1, (k1, p1) to end.
Next row K1, (p1, k3 tog) to last 2 sts, p1, k1.
Next row P1, (p2 tog) to end. 12(13,14) sts.
Break off yarn, thread end through rem sts, pull up and secure. Join seam, reversing seam on last 5cm/2in for brim. Turn back brim.

SHOES
With $3\frac{1}{4}$mm (10) needles cast on 38(42,46) sts. K 3 rows. Beg with a k row, work 6 rows in st st. K 3 rows. Work 10 rows in k1, p1 rib. Beg with a k row (thus reversing fabric), work 4 rows in st st.
Shape Instep
Next row K13(15,17), p12, turn.
Next row K12, turn.
Work on these 12 sts only.
Next row P5, k twice in each of next 2 sts, p5, 14 sts.
Work in patt as follows:
1st row (Wrong side) k5, p4, k5.
2nd row P4, C3R, C3L, p4.
3rd row K4, p2, k2, p2, k4.
4th row P3, C3R, p2, C3L, p3.
5th row K3, p2, k4, p2, k3.
6th row P2, C3R, p2, MB, p1 then pass bobble st over this st, p1, C3L, p2.
7th row K2, p2, k6, p2, k2.
8th row P2, C3L, p4, C3R, p2.
9th row As 5th row.
10th row P3, C3L, p2, C3R, p3.
11th row As 3rd row.
12th row P4, C3L, C3R, p4.
13th row K5, (p2 tog) twice, k5, 12 sts.
Beg with a p row, work 0(2,4) rows in reverse st st. Break off yarn.
Leave these sts on a holder.
With right side facing, rejoin yarn at base of instep, pick up and k10(11,12) sts evenly along side edge of instep, k across sts on holder, pick up and k10(11,12) sts evenly along other side edge of instep, k rem 13(15,17) sts. 58(64,70) sts. K 8 rows.
Next row (Pick up st 7 rows below corresponding with next st on left-hand needle and k them tog) to end.
Beg with a k row, work 8 rows in st st. Break off yarn.
Shape sole With right side facing, slip first 22(25,28) sts onto right-hand needle, rejoin yarn to rem sts and work as follows:
Next row K13, k2 tog, turn.
Next row Slip 1, k12, k2 tog tbl, turn.
Next row Slip 1, k12, k2 tog, turn.
Rep last 2 rows 8(10,12) times more, then work first of the 2 rows again.
Next row Slip 1, k4, k2 tog tbl, k2 tog, k4, k2 tog, turn.
Next row Slip 1, k10, k2 tog tbl, turn.
Next row Slip 1, k10, k2 tog, turn.
Rep last 2 rows once more, then work first of the 2 rows again.
Next row Slip 1, k3, k2 tog tbl, k2 tog, k3, k2 tog, turn.
Next row Slip 1, k8, k2 tog tbl, turn.
Next row Slip 1, k8, k2 tog, turn.
Rep last 2 rows 3(4,5) times.
Next row Slip 1, k2, k2 tog tbl, k2 tog, k2, k2 tog tbl, turn. Place rem 4 sts at each side of centre (sole) sts on one needle, pointing in the same directions as needle with sole sts.
With right side of sole and upper sts together, cast off together rem 8 sts. Join seam, reversing seam on cuff. Turn back cuff. Make one more.

Saints Design Cardigan

BY J & J SEATON

This is probably the most complicated knitting pattern in the whole book so if you are a beginner, you should polish your skills on another knit!

MEASUREMENTS

To fit bust size 91 (96,102)cm/36 (38,40)in
Actual size 95 (100,105)cm/37½(39½,41½)in
Length 57(58,59)cm/22½(23,23½)in
Sleeve seam 47(48,49)cm/(18½(19,19½)in

TENSION

The tension of this garment should be 33 st and 39½ rows to 10cm (4in) over the pattern using 2¾mm (12) needles. If you cannot obtain this using the size of needles recommended, use a larger or smaller pair accordingly.
N.B. Row numbers refer to rows worked above the ribbing.

ABBREVIATIONS

alt–alternate, **beg**–beginning, **cont**–continue, **dec**–decrease, **foll**– follow(ing), **inc**–increase, **k**–knit, **m**–make, **patt**–pattern, **p**–purl, **rem**–remaining, **rep**–repeat, **st(s)**–stitch(es), **st st**–stocking stitch, **tb st**–through back of stitch.

NOTE ON INTARSIA

All parts of the patt should be worked *strictly* in intarsia – see p. 25 for full instructions. *Even where there are only two colours on a row do not be tempted to work in Fair Isle as this will throw your tension out and spoil the feel of the garment.*

MATERIALS

You will require yarns as follows:

A	275g	Purple Lake Botany	652	RY
B	75g	Beetle Silk Silkstones	834	RY
C	100g	Wine Shetland	55	JS
D	45g	Honey Gold Paterna	730	P (S)
E	60g	Navy Shetland	120	JS
F	20g	Aquamarine Shetland	142	JS
G	20g	Olive Green Paterna	652	P (S)
H	50g	Dark Blue Botany	54	RY
J	15g	Dull Lilac Paterna	D127	P(S)
K	30g	Tawny Gold Paterna	750	P (S)
L	55g	Sand Shetland FC	45	JS
M	20g	Dull Violet Botany	118	RY
O	20g	Raw Umber Botany	80	RY
P	45g	Dove Grey Botany	59	RY
R	20g	Gentian Shetland	131	JS
S	30g	Indigo Violet Botany	527	RY
T	15g	Mustard Shetland	28	JS
V	20g	Glacier Blue Paterna	561	P (S)
W	15g	Chestnut Botany	77	RY

One pair each of 2¼mm (13) and 2¾mm (12) needles; six buttons

NOTE ON MATERIALS

RY = Rowan Yarns, P(S) = Paterna (Stonehouse), JS = Jamieson & Smith
Paterna yarn is made up of three strands loosely wound together in each ball. Cut off the length required and split off just *one* strand for use in your knitting.

Back
For charts A and B see pp. 26–29.
With $2\frac{1}{4}$mm (13) needles and colour A, cast on 147 (155,163) sts. Rib 20 rows in twisted rib as folls:
Row 1 K1 tb st * p1, k1 tb st, rep from * to end.
Row 2 P1 * k1 tb st, p1, rep from * to end.
Rep these 2 rows 8 times more. Rep row 1 again.
Inc row (20th ribbing row) in twisted rib as set, rib 2(6,10) * m1, rib 9 rpt from * ending last rpt rib 1(5,9) to 164(172,180) sts.
Change to $2\frac{3}{4}$mm (12) needles. Now work on as shown on the back chart, working in st st and patt as shown, beg with a k row.
Work on, in st st and patt as set to row 104(106,110).

Shape armholes Cast off 5 sts at beg of next 2 rows to 154(162,170 sts). Work on to row 197(201,207).
Divide sts for back neckline: next row: p61(64,67) and leave on a spare needle until required for left back shoulder. Cast off central 32(34,36) sts for back of neck. P to end. Cont on these 61(64,67) sts for right back shoulder.
Right back shoulder Slope shoulder and shape back neckline: work 1 row.
Row 200(204,210) Cast off 8, work to end of row.
Row 201(205,211) Cast off 17(18,19), work to end of row, dec 1.

Row 202(206,212) Dec 1, work to end of row.
Row 203(207,213) Cast off 17(18,19), work to end of row, dec 1.
Row 204(208,214) Work 1 row on rem 16(17,18) sts.
Cast off.
Left back shoulder With right side facing, rejoin yarn to inner edge of 61(64,67) sts left on spare needle and work in patt as set to end of row.
Now work as given for right back shoulder to end.

Left Front
For charts C and D see pp 30–33.
With $2\frac{1}{4}$mm (13) needles and colour A, cast on 81(85,89) sts. Rib 20 rows in twisted rib as given for back, *except* for the 10 sts on the inside edge of the piece, which should be worked in ordinary single rib. These 10 sts form the base of the buttonband, which will continue up the piece as far as the shoulder and then sufficient rows further to form the back of neckband.
Inc row (20th ribbing row) work 10 sts in single rib for the buttonband, then in twisted rib as set, rib 4(6,8) * ml, rib 8, rep from * ending last rep rib 3(5,7) to 90(94,98) sts.
Change to $2\frac{3}{4}$mm (12) needles. Now work as shown on left front graph chart. Beg with a k row and work on in st st – except for the 10 sts on the inside edge of the piece which should be worked in single rib to make the buttonband – and patt as shown. All parts of the patt should be worked strictly in intarsia. Work on, in rib, st st and patt set to row 104(106,110).
Shape armholes Cast off 5 sts at beg of next row to 85(89,93) sts. Work on to row 147(149,153).
Shape V-neck In order not to interfere with the front band, all the V-neck decreases are fully-fashioned and are worked within the st st part of the garment, as folls:
Next row Work 10 sts in single rib for front band as set, p next 2 sts tog, p to end in patt as set.
Rep this row in foll 24(25,26) alt rows to 60(63,66) sts and row 196(200,206). Work on to row 202(206,212).
Slope shoulder Cast off 17(18,19) sts at beg of next row and foll alt row, then cast off 16(17,18) sts at beg of next alt row. This leaves the 10 sts of ribbing for the front band. To provide a band for the back of the neck continue to work on these 10 sts, in ribbing as set, until you have a sufficient length. Leave these 10 sts on a spare piece of yarn or a safety pin.

Right Front
Work the right front as given for left front, but reverse all shapings and follow the right front graph chart (see pp 30–33).
Also, the buttonband on the right front requires 6 buttonholes, work these as folls:
On the rib Rib 4(4,6) rows as set.
Row 5(5,7) (First buttonhole row) in single rib, rib 4, cast off 2, rib 4, work to end of row in twisted rib.
Row 6(6,8) (Second buttonhole row) in twisted rib, rib 71(75,79), then in single rib, rib 4, turn, cast on 2, turn, rib to end. Work 14 rows more in rib as set, remembering to inc on row 20 thus: in twisted rib as set, rib 4(6,8) * m1 rib 8, rep from * ending last rep rib 3(5,7) in twisted rib, rib 10 in single rib – to 90(94,98) sts.
Change to $2\frac{3}{4}$mm (12) needles. Work on as shown on right front graph chart, beg with a k row and working in st st – except for the 10 sts of ribbing forming the buttonhole band on the front edge – and patt as set.
Make buttonholes Work on as shown on the chart to row 16(16,18).
Row 17(17,19) (First buttonhole row) rib 4, cast off 2, rib 4, k in patt as set to end of row.
Row 18(18,20) (Second buttonhole row) p80(84,88) in patt as set, rib 4, turn, cast on 2, turn, rib to end.
* Work 30 rows as given on chart, rep row 17(17,19) and row 18(18,20).
Rep from * 3 times more. This gives buttonholes on row 17(17,19), row 49(49,51), row 81(81,83), row 113(113,115) and row 145(145,147).
At the same time, shape armholes Work to row 105(107,111): cast off 5 sts at beg of next row – 85(89,93)sts.
Work to row 147(149,153).
Shape V-neck As for the left front, all V-neck dec are worked within the st st part of the piece, as folls:
Next row P73(77,81) in patt as set, p2 sts tog tb st, work 10 sts in single rib for the frontband. Rep

this row on foll 24(25,26) alt rows to 60(63,66) sts and row 196(200,206).

Slope shoulder Cast off 17(18,19) sts at beg of next row and foll alt row. Work 1 row on rem 26(27,28) sts. Cast off 16(17,18) sts of st st, leaving the rem 10 sts of ribbing on a spare piece of yarn or a safety pin.

Sleeves

For charts E and F see pp. 34–37.

With $2\frac{1}{4}$mm (13) needles and colour A, cast on 69(71,73) sts. Rib 24 rows in twisted rib as given for back.

Inc row (24th Ribbing row) in twisted rib as set, rib 4(6,6) * m1, rib 3, repeat from * ending last repeat rib 5(5,7) to 90(92,94) sts. Change to $2\frac{3}{4}$mm (12) needles. Now work on as shown on charts E and F, pp. 34–37, beg with a k row and working in st st and patt as set. At the same time, inc 1 st at each end of row 4 and every foll 4th row 36(37,38) times more to 164(168,172) sts and row 148(152,156). Work on to row 172(176,180). Cast off.

Finishing

Join shoulder seams and set in sleeves. Join sleeve and side seams. For the neck ribbing, join the extra length knitted on buttonband to the back neck edge with a flat seam and graft together the two groups of stitches from the buttonband and buttonhole band which have been left on a length of yarn or safety pin. Sew on buttons. Steam press on wrong side only. Finish ends securely and neatly. Take great care not to cut these ends too short or they will wriggle loose with wear. A good 4cm ($1\frac{1}{2}$in) should be worked firmly into the backs of the motifs (preferably using a latchet tool) rather than into the surrounding area of background, where they will tend to show through.

SPECIAL INTARSIA INSTRUCTIONS

All Seaton designs should be worked by the intarsia method. This is a particular way of handling the yarns and should not be confused with Fair Isle. Fair Isle is a method of patterning where the yarns are carried across the back which produces a fabric of double thickness. This method should never be used on Seaton garments as, with the amount of colours used, it would produce an impossibly thick and inflexible garment!

To work intarsia correctly, each colour is knitted with a separate length of yarn and the different colours are twisted together at each colour join and then left in position to be used on the next row. They are *not* carried across the back.

When many colours are being worked together in one garment, it can be difficult to avoid a dreadful tangle at the back of the work. It is much better to cut off lengths of yarn rather than working straight off the ball so you can pull the length free of the rest of the yarns in use. It is usually quite possible to estimate the length of yarn needed for each area so there is little wastage.

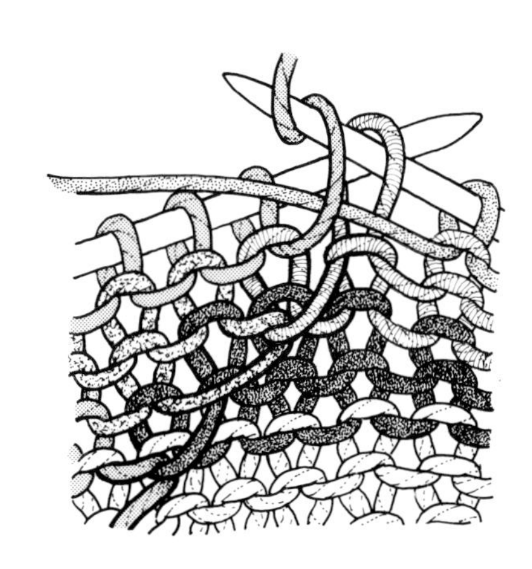

46" (102cm)
38" (96cm)
36" (91cm)

36"
(91CM)
38"
(96CM)
40"
(102CM)

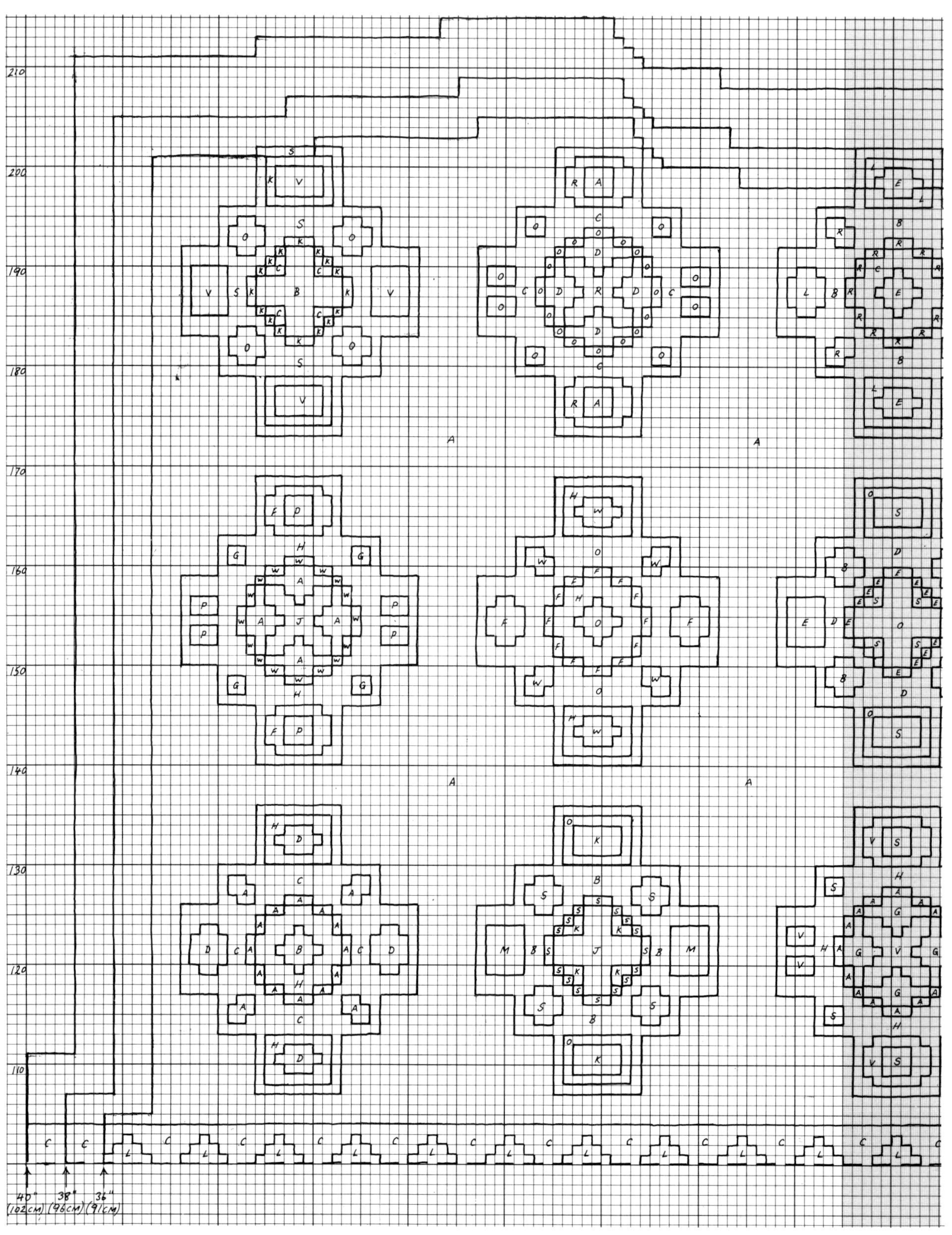
210
200
190
180
170
160
150
140
130
120
110
40" (102cm)
38" (96cm)
36" (91cm)

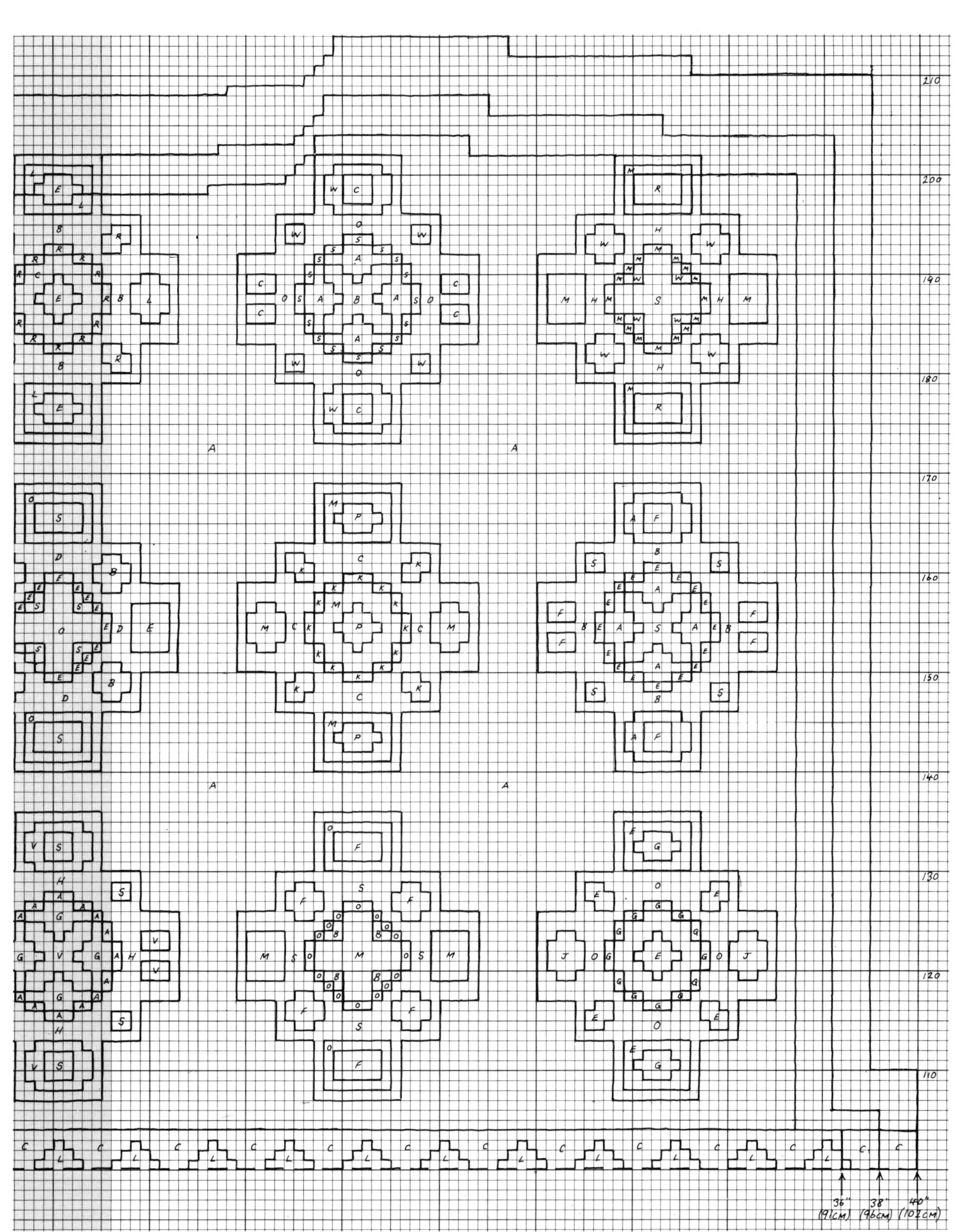
210
200
190
180
170
160
150
140
130
120
110
36" (91CM)
38" (96CM)
40" (102CM)

CHART C – LEFT SIDE, FRONT, BOTTOM

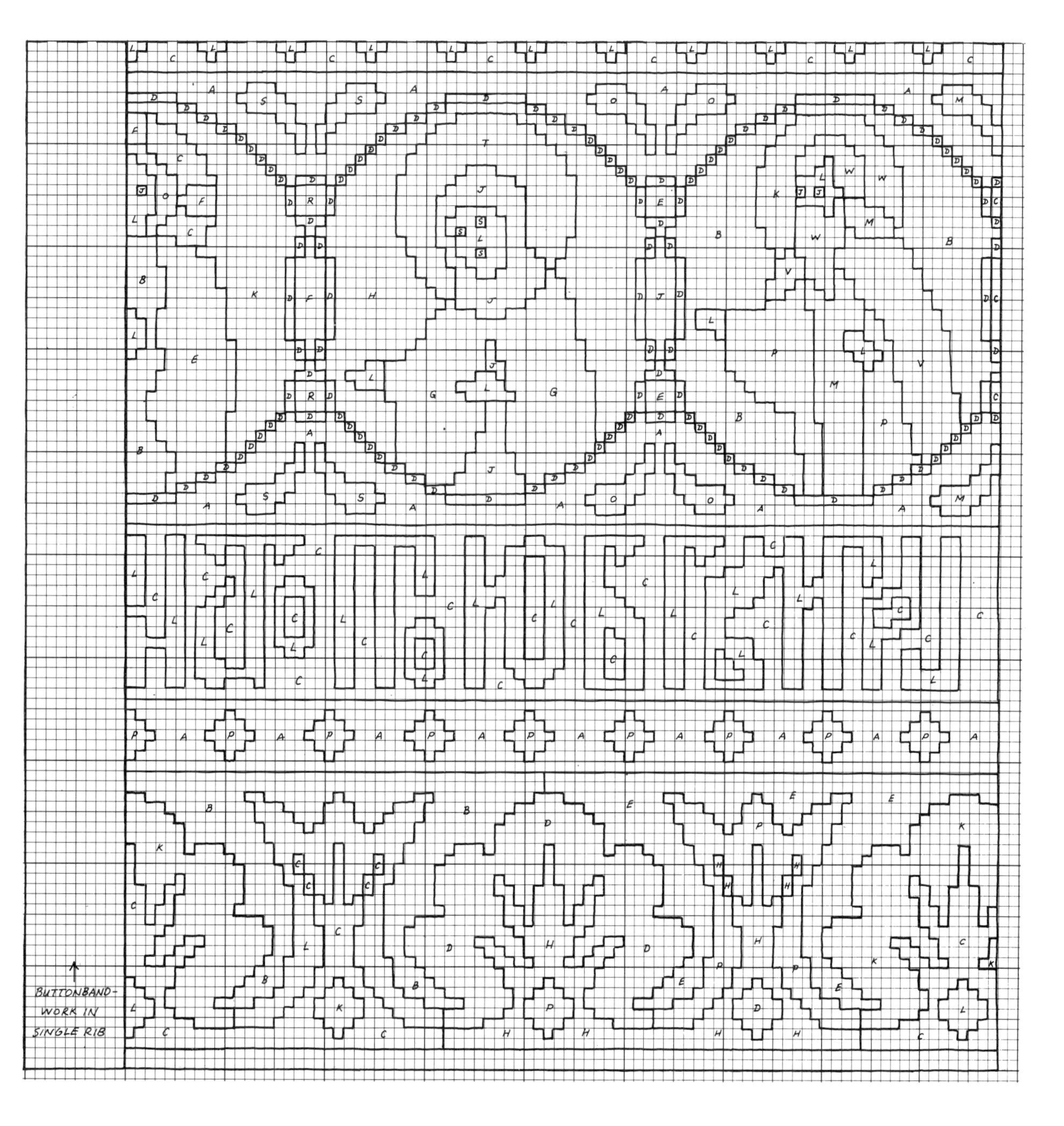
BUTTONBAND-
WORK IN
SINGLE RIB

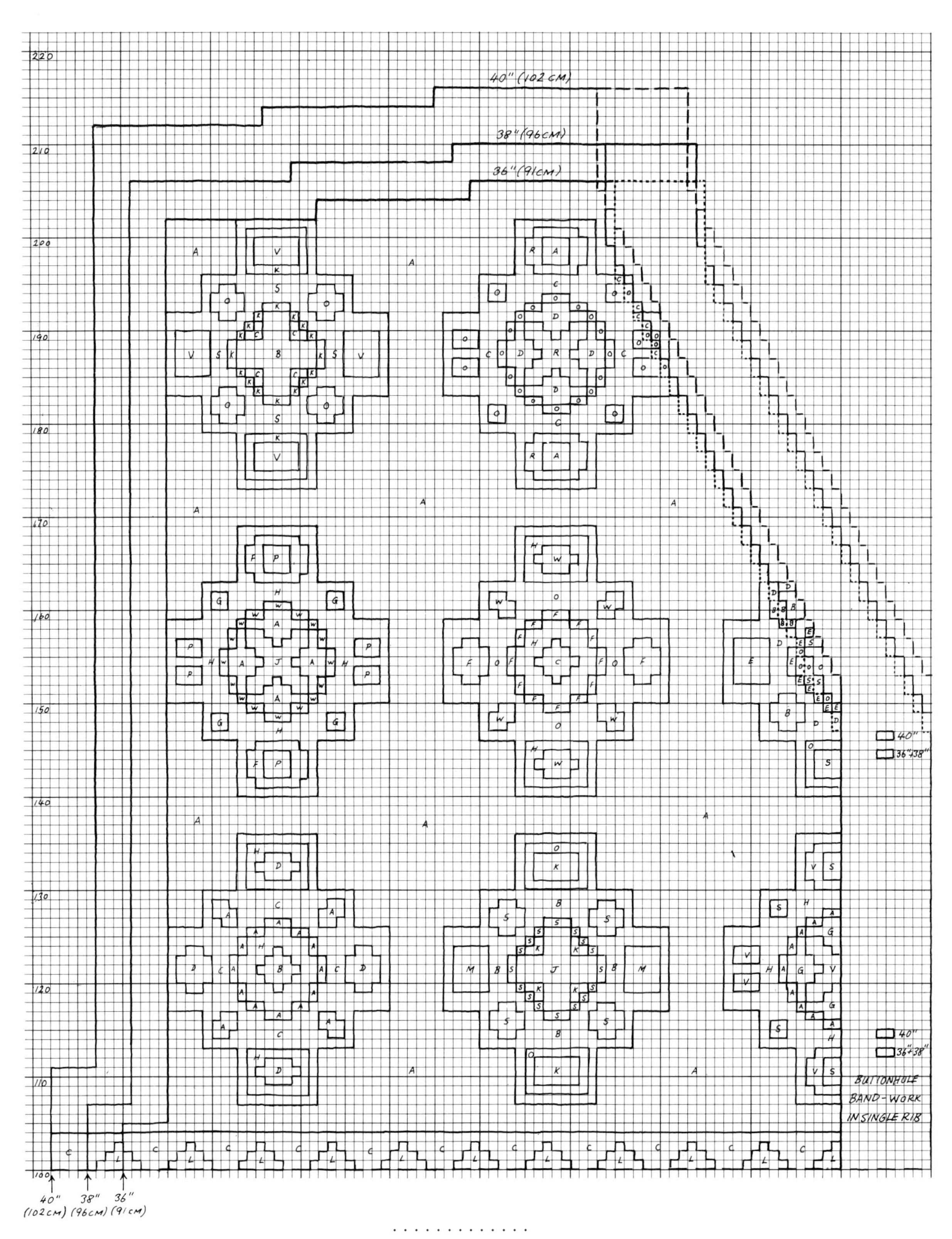
40" (102 CM)
38" (96 CM)
36" (91 CM)
220
210
200
190
180
170
160
150
140
130
120
110
100
40"
36"-38"
BUTTONHOLE
BAND-WORK
IN SINGLE RIB
40" (102CM)
38" (96CM)
36" (91CM)

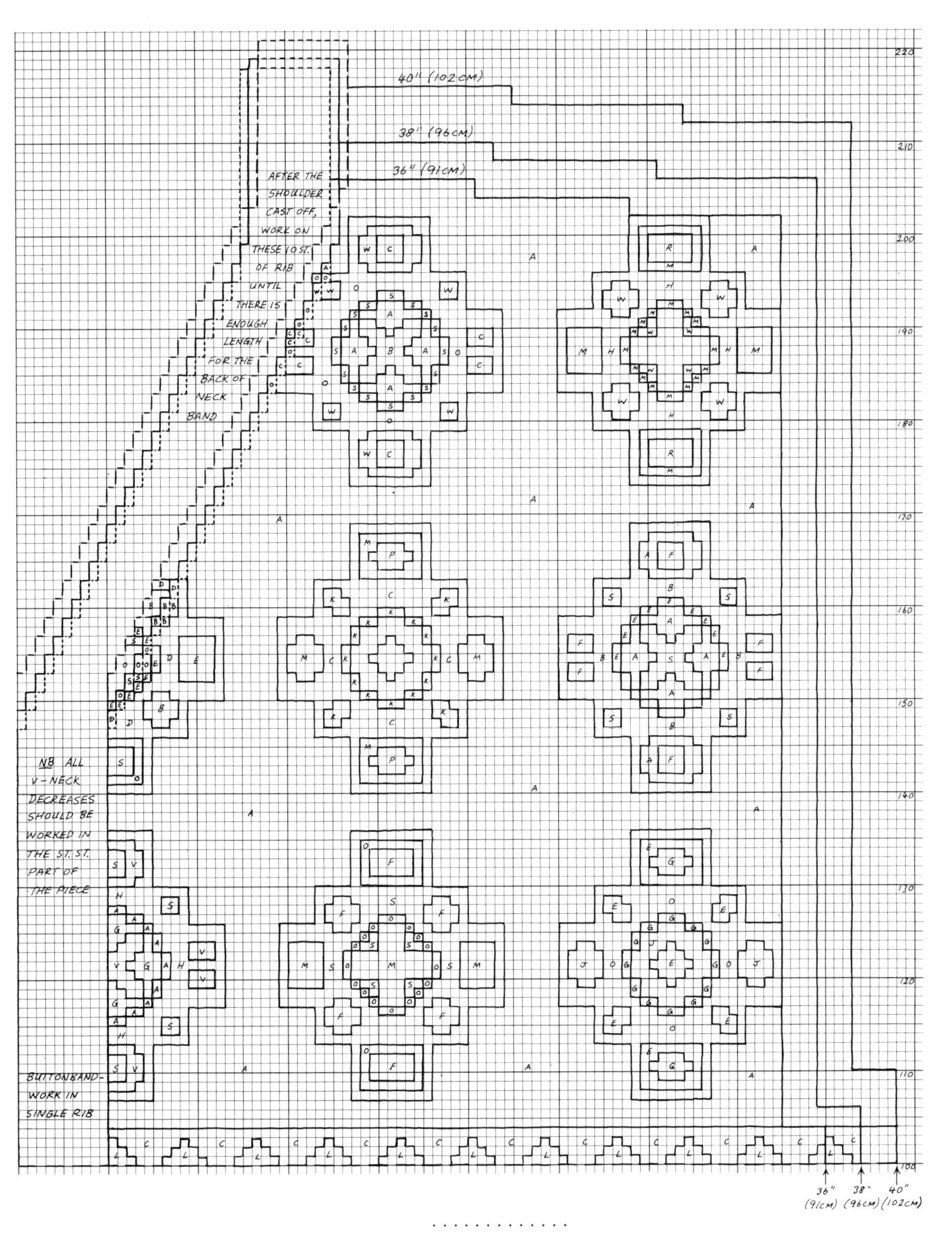

40" (102 CM)
38" (96 CM)
36" (91 CM)
AFTER THE SHOULDER CAST OFF, WORK ON THESE 10 ST. OF RIB UNTIL THERE IS ENOUGH LENGTH FOR THE BACK OF NECK BAND
NB ALL V-NECK DECREASES SHOULD BE WORKED IN THE ST. ST. PART OF THE PIECE
BUTTONBAND-WORK IN SINGLE RIB
220
210
200
190
180
170
160
150
140
130
120
110
100
36" (91CM)
38" (96CM)
40" (102CM)

100
90
80
70
60
50
40
30
20
10
0

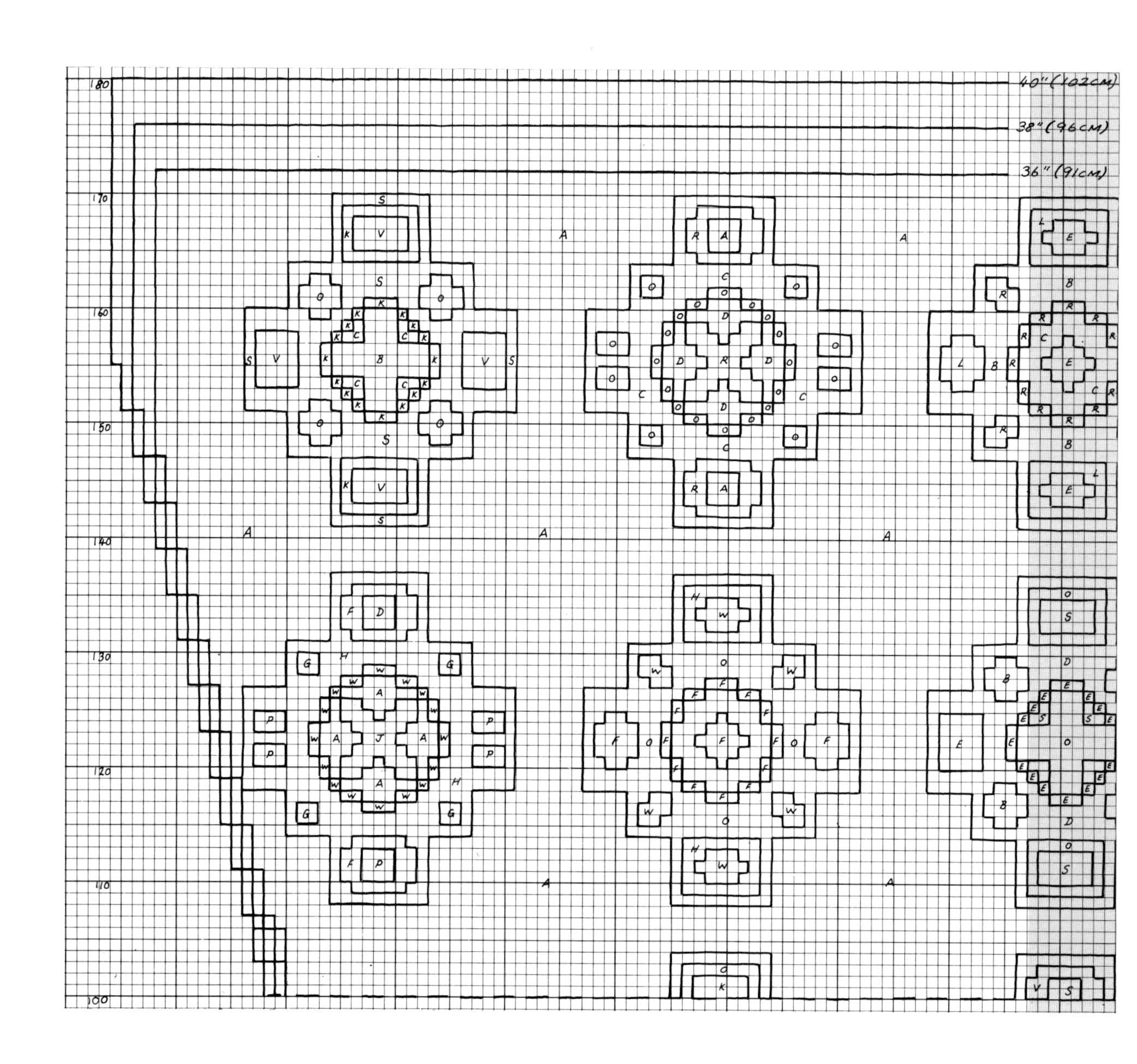
40" (102CM)
38" (96CM)
36" (91CM)
180
170
160
150
140
130
120
110
100

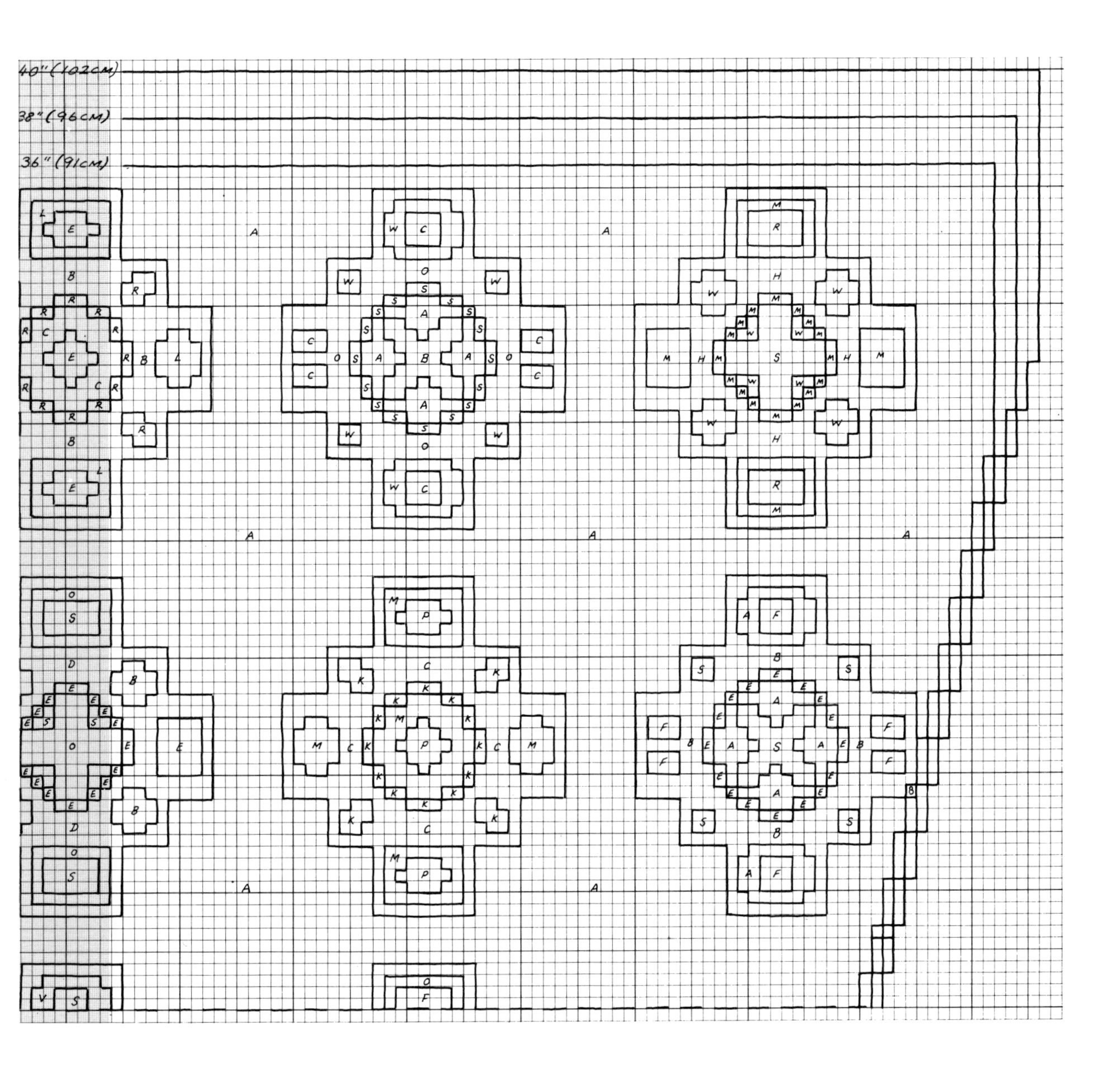
40" (102CM)
38" (96CM)
36" (91CM)

Postman Pat Jumper

BY GARY KENNEDY

This quick and easy sweater will delight children of all ages and perhaps some adults too! Worked in double knitting and using the intarsia method for colours, this pattern needs good tension and some experience.

MEASUREMENTS

Chest	Actual	Length	Sleeve seam
cm(in)	*cm(in)*	*cm(in)*	*cm(in)*
56(22)	66(26)	38(15)	25(10)
61(24)	71(28)	46(18)	30(12)
66(26)	76(30)	48(19)	33(13)
71(28)	81(32)	48(19)	38(15)
76–81(30–32)	96(38)	61(24)	46(18)
86–91(34–36)	107(42)	64(25)	48(19)
96–102(38–40	117(46)	66(26)	48(19)

MATERIALS
4(5,5,6,8,10,11) 50 gram balls DK in MC (Main Colour). One ball of each other colour (see p. 40).
Pair each of 3mm (11) and 3¾mm (9) needles
Spare stitch holder

TENSION
6 sts and 8 rows = 2.5cm (1in) on 3¾mm (9) needles. Check your tension, if less sts, use a finer needle, if more sts, use a coarser needle.

ABBREVIATIONS
alt–alternate, **beg**–begin(ning), **cont**–continue, **foll**–follow(ing), **dec**–decrease, **inc**–increase, **k**–knit, **MC**–main colour, **patt**–pattern, **p**–purl, **rem**–remaining, **rep**–repeat, **sl**–slip, **st(s)**–stitch(es), **st st**–stocking stitch, **tog**–together.
N.B. Instructions for the larger sizes are in brackets; where one figure is given, this applies to all sizes.

METHOD OF WEAVING (INTARSIA)
When working chart pattern use a separate ball of wool for each section, twisting them together to avoid a hole when changing colour. (See p. 25).

Front
**With MC and 3mm (11) needles cast on 71(75,81,87,101,113,125) sts
1st row (Right side) k1, *p1, k1 Rep from * to end of row.
2nd row P1, *k1, p1. Rep from * to end of row.
Repeat these two rows (k1, p1) ribbing for 2.5(5,5,5,7.5,7.5,7.5)cm/1(2,2,2,3,3,3)in, ending with right side facing for next row and increasing 9(9,9,9,13,13,13) sts evenly across last row. 80(84,90,96,114,126,138) sts on needle.**
Change to 3¾mm (9) needles.
1st size only After rib, beg working grid (see chart on p. 41) immediately from row 6. Cont working with grid, end on row 105. Shape neck.
2nd to 7th sizes Beg with a k row k2(2,4,24,24,28) rows in st st.
Next row K2(5,8,17,23,29) sts. Knit 80 (i.e. total width of grid) to end. K2(5,8,17,23,29) sts. Total 84(90,96,114,126,138) sts. This sets position of grid.
Work in st st with 80 sts across a 110 row grid until 116(116,116,144,148,156) rows have been worked in total from beg.

Shape Neck
Next row (Right side) k30(32,34,37,46,50,56) sts (neck edge), turn. Leave rem sts on spare needle. Dec 1st at neck edge on next 6(6,6,7,8,8,9) rows. 24(26,28,30,38,42,47) sts on. Work straight in st st for 9(9,9,12,15,19,18) rows ending with right side facing for next row.

To Shape Shoulder
Cast off 8(9,9,10,12,14,16) sts at the beg of the next and foll alt row. Work 1 row even. Cast off

PAT

rem 8(8,10,10,14,14,15) sts. With right side facing, slip next 20(20,22,22,22,26,26) sts from spare needle onto a st holder. Join MC to rem sts and knit to end of row. Work to correspond to other side, reversing all shapings.

Back
Work from ** to ** as given for front. Change to $3\frac{3}{4}$mm (9) needles. Proceed in st st until back measures same length as front to shoulder shaping, ending with right side facing for next row.

To Shape Shoulder
Cast off 8(9,9,10,12,14,16) sts at the beg of the next 4 rows, then 8(8,10,10,14,14,15) sts at the beg of foll 2 rows. Leave rem 32(32,34,36,38,42,44) sts on a st holder.

Sleeves
With MC and 3mm (11) needles, cast on 35(45,45,49,53,63,69) sts and work 2.5(5,5,5,7.5,7.5,7.5)cm/1(2,2,2,3,3,3)in in k1, p1 rib as given for front, increasing 5(7,7,9,9,9,9) sts evenly across last row to 40(52,52,58,62,72,78) sts on needle. Change to $3\frac{3}{4}$mm (9) needles, proceed in st st inc 1 st at each end of needle on 3rd, then every foll 4th row 13(0,6,2,11,3,3) times to 68(54,66,64,86,80,86) sts on needle. Work 5 rows even. Inc 1 st at each end of needle on next, then every foll 6th row until there are 72(78,84,90,108,114,120) sts on needle. Cont even in st st until sleeve measures 25(30,33,38,46,48,48)cm/10(12,13,15,18,19, 19)in from beg. Cast off loosely.

To Make Up and Neckband
Sew right shoulder seam. With MC, 3mm (11) needles and right side of work facing, pick up and k 19(19,19,23,27,31,31) sts down left front neck edge, k across 20(20,22,22,22,26,26) sts from front st holder. Pick up and knit 19(19,19,23,27,31,31) sts up right front neck edge. Knit across 32(32,43,36,38,42,44) sts from back st holder dec 1 st in centre. 89(89,93,103,113,129,131) sts on needle. Work in k1, p1, rib for 5–7.5cm (2–3in). Cast off loosely.
Sew left shoulder and neckband seam. Fold neckband in half to wrong side and sew loosely in position. Place markers on front and back side edges 15(16.5,18,19,23,23,25)cm/6($6\frac{1}{2}$,7,$7\frac{1}{2}$,9, 9,10)in down from shoulder seams. Sew in sleeves between markers. Sew side and sleeve seams. Omitting ribbing, press lightly under a dry cloth using a cool iron.

COLOUR KEY

A	White	E	Yellow
B	Blue	F	Red
C	Flesh	G	Black
D	Rust	MC	Main Colour

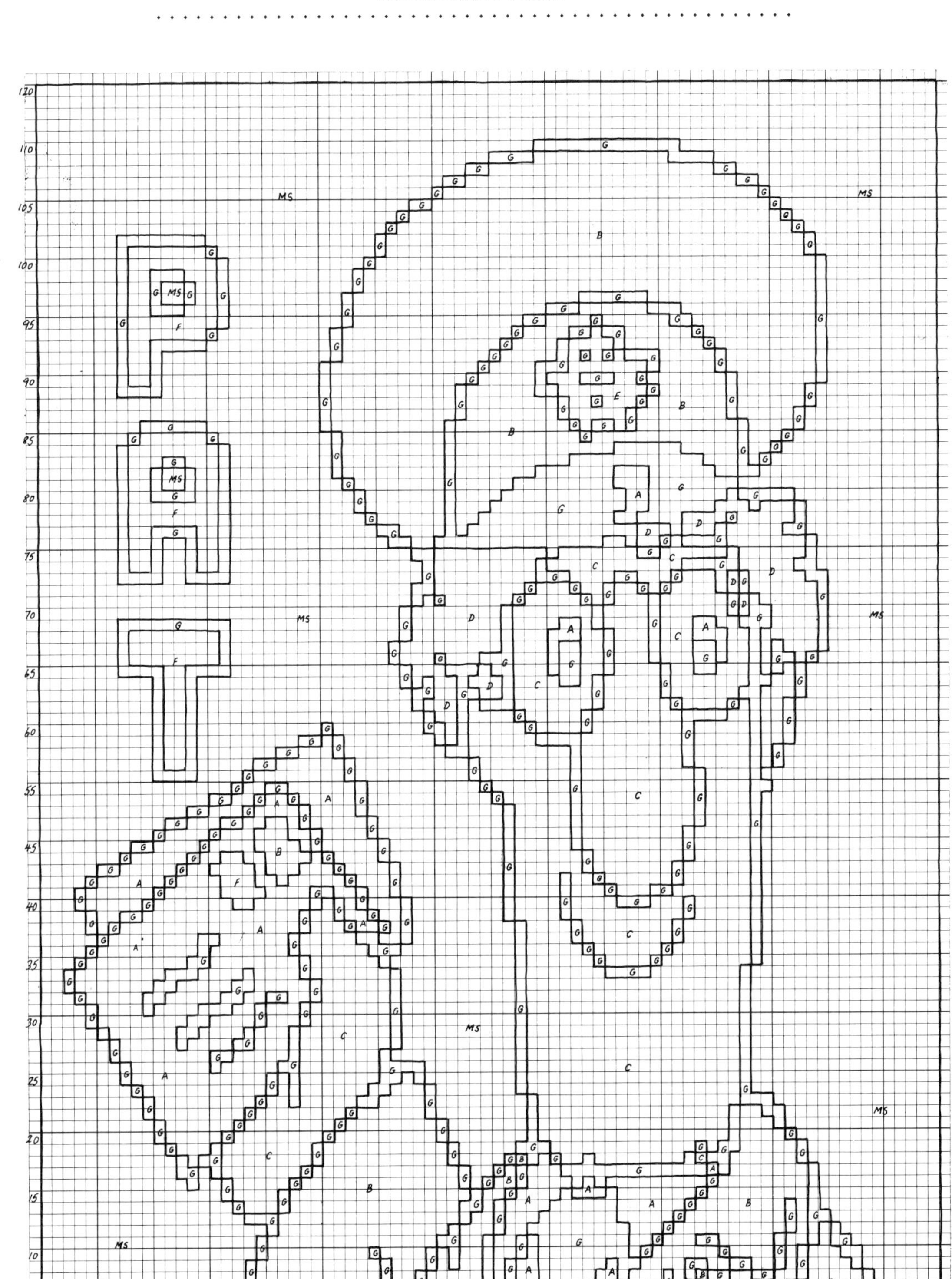
120
110
105
100
95
90
85
80
75
70
65
60
55
45
40
35
30
25
20
15
10
5
0
5
10
15
20
25
30
35
40
45
50
55
60
65
70
75
80

Irish Moiled Jacket

BY LESLEY CONROY

This jacket combines intarsia work with cabling to make another fairly challenging pattern. The natural wools are really lovely to work with because they leave your hands beautifully soft and oily.

MEASUREMENTS

One size only
Actual 122cm (48in)
Length 69cm (27in)
Sleeve length 51cm (20in)
N.B. If a larger size jacket is desired it is possible to work the instructions as given but substituting 4mm (8) needles for $3\frac{3}{4}$mm (9) needles and 5mm (6) needles for $4\frac{1}{2}$mm (7) needles. This will give a jacket of approximately 132cm(52in) in width.

MATERIALS

Wools available in kit form from Lesley Conroy or direct from British Mohair Spinners, Bradford.

A	800g	Cream	Aran
B	300g	Mid-Grey Welsh	British
C	200g	Black Welsh	British

Small amounts of the following are needed for horns:

D	——	Dark Grey Welsh	British
E	——	Moorit	Shetland

One pair each of $3\frac{3}{4}$mm (9) and $4\frac{1}{2}$mm (7) needles
1 cable needle
Spare stitch holders
10 buttons

TENSION

20 sts and 26 rows for a 10cm (4in) square using $4\frac{1}{2}$mm (7) needles over patterned stocking stitch.
25 sts and 28 rows for a 10cm (4in) square using $4\frac{1}{2}$mm (7) needles over cable and bobble pattern.
(For larger size: 18 sts and 24 rows for a 10cm (4in) square using 5mm (6) needles over stocking stitch).

If too few stitches to 10cm (4in), use a finer needle; if too many, use a coarser needle. It is important to check your tension before you start to knit. Knit a tension square and change your needle size if necessary.

ABBREVIATIONS

alt–alternative, **beg**–begin(ning), **cn**–cable needle, **cont**–continue, **dec**–decrease, **foll**–follow(ing), **inc**–increase, **k**–knit, **k1b**–knit one stitch through back loop, **k2tog b**–knit 2 together through back loop, **k3 tog**–knit 3 sts together, **MC**–main colour, **patt**–pattern, **p**–purl, **p1b**–purl one st through the back loop, **psso**–pass slipped st over, **rem**–remaining, **rep**–repeat, **st(s)**–stitch(es), **st st**–stocking stitch, **tog**–together

C4B–(Cable 4 Back) work as follows: place 2 sts on cable needle and leave at back of work, k2, then k2 from cable needle.

C4F–(Cable 4 Front) work as follows: place 2 sts on cable needle and leave at front of work, k2, then k2 from cable needle.

C6B–(Cable 6 Back) work as follows: place 3 sts on cable needle and leave at back of work, k3, then k3 from cable needle.

C6F–(Cable 6 Front) work as follows: place 3 sts on cable needle and leave at front of work, k3, then k3 from cable needle.

LT–(Left twist) work as follows: with right-hand needle behind left-hand needle, miss one st and k the second st in back loop; then insert right-hand needle into the back of both sts (the missed st and the second st) and k2tog through back loops.

MB–(Make bobble) see Stitch Notes page 30.

RT–(Right twist) work as follows: k2tog, leaving

st on left-hand needle; then insert right-hand needle from the front between the two sts just knitted together, and knit the first stitch again; then slip both stitches from the needle together.

NOTES

1 Charts A, B, C, D and E are worked throughout in stocking stitch, starting with a knit row. Right side (odd numbered) rows are worked from right to left and wrong side (even numbered) rows are worked left to right.
2 Rows 1–9, 13–19 and 49–56 of all the charts are worked using the Fair Isle method, i.e. both yarns are carried across the back of the work by stranding and weaving in every 3rd stitch where necessary to avoid long floats. Do not use the intarsia method for these rows as it would impair the stability of the fabric.
3 The cow motifs are worked by the intarsia method, using separate lengths of yarn for each colour and twisting the yarns around each other where they meet on the wrong side to avoid holes. The background yarn may be carried loosely across the back of the work, weaving in where necessary.
4 The Cream and Moorit and Dark Grey Welsh details on the cows may be either knitted in or Swiss darned afterwards if preferred.
5 The cows' bodies are worked mainly in Black Welsh.
6 Make increases on sleeves on 3rd stitch from each end using the Make 1 method, i.e. insert needle into horizontal strand lying between stitch and knit through back loop to twist the stitch and thus prevent a hole.
7 Leave long lengths of yarn at all cast-on and cast-off edges so that they can be used for sewing up. This creates a more professional finish.
8 Weave in all ends of colours carefully after motifs have been completed so that there is no possibility of yarn unravelling.
9 The button band and buttonhole bands are worked at the same time as the fronts in moss stitch. Please note carefully when to change colour when working the bands – it is indicated on the front charts and in the instructions. Do not forget to make the buttonholes on the buttonhole band to correspond with button

placement markers. See instructions for further details.

STITCH NOTES

Moss stitch Worked over 5 sts for button band and buttonhole band.
All rows *K1, p1, rep from * ending k1.
Bobble Pattern Worked over 3 sts.
Row 1 (Right side) k3.
Rows 2, 4 and 6 P3.
Row 3 K1, make bobble (MB) in next st, k1.
Rows 5 and 7 K3.
Row 8 P3.
Repeat rows 1 to 8 to form the pattern.
To make bobble – work as follows:
knit in front and back of next stitch twice, then knit in front again, (turn, p5, turn, k5) twice, pass 2nd, 3rd, 4th and 5th stitches one at a time over the first stitch and off the needle to complete the bobble.

Cable pattern Worked over 12 stitches.
Row 1 (Right side) p2, k8, p2.
Row 2 and all alternate wrong side rows K2, p8, k2.
Row 3 P2, C4B, C4F, p2.
Row 5 P2, k8, p2.
Row 7 P2, k2, C6B, p2.
Row 9 P2, k8, p2
Row 11 P2, C6F, k2, p2.
Row 13 P2, k8, p2.
Row 15 P2, C4F, C4B, p2.
Row 16 K2, p8, k2.
Rep rows 1 to 16 to form cable pattern.

Back
Using $3\frac{3}{4}$mm (9) needles and Black Welsh yarn, cast on 107 sts. Work in bobble rib as follows:
Row 1 (Right side) *p2, k3, rep from * to end, ending p2.
Rows 2, 4 and 6 *K2, p3, rep from * ending k2.
Row 3 *P2, k1, MB in next st, k1, p2, k3, rep from * to last 7 sts, ending p2, k1, MB, k1, p2.
Rows 5 and 7 As row 1.
Row 8 *K2, p3, rep from * ending k2.
Rows 1–8 form the bobble rib patt. Work a total of 28 rows of rib pattern increasing 13 sts evenly over the last row, 120 sts.
Change to $4\frac{1}{2}$mm (7) needles and, joining in different colours as required, work the 56 rows of Chart A entirely in st st, starting with a k row. Rows 1–9, rows 13–19 and rows 49–56 are worked using the Fair Isle technique, i.e. both yarns are carried across the back of the work by stranding and weaving in every 3rd stitch where necessary. Do not work the intarsia method for these rows as it would impair the stability of the fabric. The cow motifs are worked using the intarsia method, joining in separate lengths or bobbins of yarn for each cow motif and linking one colour to the next by twisting them around each other where they meet on the wrong side to avoid holes. The background yarn, Mid-Grey Welsh, should be carried right across the back of the work, weaving in where necessary to avoid long floats.
When the 56 rows of Chart A have been completed, fasten off Mid-Grey Welsh yarn and continue in Cream Aran only.
Next row K, increasing 35 sts evenly across the row, 155 sts.
Next row P.
Now begin to work in cable and bobble patt as folls:
Row 1 (Right side) k1, * work row 1 of bobble patt over next 3 sts, work row 1 of cable patt over next 12 sts, rep from * to last 4 sts, work row 1 of bobble patt over next 3 sts, k1.
Row 2 P1, * work row 2 of bobble patt over next 3 sts, work row 2 of cable patt over next 12 sts, rep from * to last 4 sts, work row 2 of bobble patt over next 3 sts, p1.
These 2 rows set the patt. Continue to work in patt as established, working appropriate rows of the bobble and cable patts. (N.B. Remember the bobble patt has an 8 row repeat and the cable patt a 16 row repeat.) Work in patt until back measures 41cm (16in) from cast-on edge, ending with a wrong side row.
Shape armholes Keeping continuity of patt cast off 16 sts at beg of next two rows, 123 sts.
Continue straight in established patt without further shaping until back measures 69cm (27in) from cast-on edge, ending with a wrong side row. Then place 37 sts on a stitch holder for shoulder seam, place 49 sts on a holder for back neck and place rem 37 sts on a holder for second shoulder seam.

Left Front
Using $3\frac{3}{4}$mm (9) needles and Black Welsh Yarn, cast on 62 sts. Work 57 sts in bobble rib as given for back, place marker, work 5 sts in moss stitch for button band.
Work a total of 28 rows in bobble rib and 5 sts moss stitch band, increasing 3 sts evenly on last row over the 57 st body section. 65 sts (i.e. 5 sts for button band and 60 sts for body section).
Change to $4\frac{1}{2}$mm (7) needles and work the 56 rows of Chart B for left front, working the colour techniques as described for the back. Continue to work the 5 sts in moss stitch for the button band in the colours indicated on the chart (i.e. work rows 1–9 of moss stitch in Black Welsh, change to Mid-Grey Welsh in row 10 and work rows 10–56 of the band in Mid-Grey Welsh).
When the 56 rows of Chart B have been

completed, fasten off Mid-Grey Welsh and continue in Cream Aran only.

Next row (Right side) knit, inc 16 sts evenly over the first 60 body sts, moss stitch 5 for the button band, 81 sts.

Next row moss stitch 5, p to end.

Now begin to work in cable and bobble patt as folls:

Row 1 (Right side) k1, * work row 1 of bobble patt over next 3 sts, work row 1 of cable patt over next 12 sts, rep from * to last 5 button band sts, moss stitch 5.

Row 2 Moss stitch 5, * work row 2 of cable patt over next 12 sts, work row 2 of bobble patt over next 3 sts, rep from * to last st, p1. These 2 rows set the patt. Cont to work in patt as established, working appropriate rows of bobble and cable patts. Work in patt until front measures same as back (41cm/16in) to armhole, ending with a wrong side row.

Shape armhole Cast off 16 sts at beg of next row, 65 sts. Now cont straight in established patt until front measures 61cm (24in) from cast-on edge, ending at neck edge.

Shape neck Keep continuity of patt as far as possible. Cast off 2 sts, work next 15 sts and place on a holder for front neck, work in patt to end, 48 sts.

Dec 1 st at neck edge on every row 7 times. Then dec 1 st at neck edge on every alt row 4 times, 37 sts.

Cont straight in patt until front measures same as back to shoulder.

Place 37 sts on a holder for shoulder seam. Mark placement for 10 buttons along button band, marking for first button 1cm ($\frac{1}{2}$in) from bottom edge and for top button 1cm ($\frac{1}{2}$in) from top of button band. Space the rest of the markers evenly between. (About every 5–6cm/2–$2\frac{1}{2}$in is a good spacing.)

Right Front

Using $3\frac{3}{4}$mm (9) needles and Black Welsh Yarn, cast on 62 sts. Work 5 sts in moss stitch for buttonhole band, work 57 sts in bobble rib as given for back.

Work a total of 28 rows in bobble rib and moss stitch buttonhole band, working first buttonhole on row 3 as folls:

Moss stitch 2, cast off next 2 sts, moss 1, bobble rib to end.

Next row Patt to where sts were cast off on previous row, cast on 2 sts. Moss to end.

Work other buttonholes in same way to correspond with markers on buttonband.

Change to $4\frac{1}{2}$mm (7) needles and work the 56 rows of Chart C for right front. When the 56 rows of Chart C have been completed, fasten off Mid-Grey Welsh and cont in Cream Aran only.

Next row Moss stitch 5, k to end increasing 16 sts evenly over the 60 body sts, 81 sts,

Next row P to last 5 sts, moss 5.

Now beg to work in cable and bobble patt as folls:

Row 1 (Right side) moss 5, * work row 1 of cable patt over next 12 sts, work row 1 of bobble patt over next 3 sts, rep from * to last 4 sts, work row 1 of bobble patt over next 3 sts, k1.

Row 2 P1, * work row 2 of bobble patt over next 3 sts, work row 2 of cable patt over next 12 sts, rep from * to last 5 sts, moss 5. These 2 rows set the patt. Cont to work in patt as established and complete right front to match left front, reversing all shapings and remembering to work the buttonholes in the appropriate places.

Right Sleeve

Using $3\frac{3}{4}$mm (9) needles and Black Welsh yarn, cast on 47 sts. Work 28 rows in bobble rib as given for back, inc 18 sts evenly over last row, 65 sts.

Change to $4\frac{1}{2}$mm (7) needles. Joining in colours as required, and working colour techniques as for back, work the 56 rows of Chart D for right sleeve, inc 1 st at each end of 3rd and every following 4th row as indicated on chart. (N.B. Try to make incs on 3rd stitch from each end where possible by the Make 1 method.)

When 56 rows of Chart D have been completed, fasten off Mid-Grey Welsh yarn and cont in Cream Aran only, 93 sts.

Next row K, inc 17 sts evenly across the row, 110 sts.

Next row P.

Now work in cable and bobble patt as folls:

Row 1 (Right side) k1, *work row 1 of bobble patt over next 3 sts, work row 1 of cable patt over

next 12 sts, rep from * to last 4 sts, work row 1 of bobble patt over next 3 sts, k1.
Row 2 P1, *work row 2 of bobble patt over next 3 sts, work row 2 of cable patt over next 12 sts, rep from * to last 4 sts, work row 2 of bobble patt over next 3 sts, p1.
These 2 rows set the patt. Now work in cable and bobble patt as established, working appropriate rows of cable and bobble patts, and inc 1 st at each of next and every alt row until there are 140 sts. Take the extra sts into the patt as they become available.
Then cont straight in patt until sleeve measures 51cm (20in) or desired length from cast-on edge. (N.B. Remember that the last 7.5cm/2¾in at the top of the sleeve fit into the cast off sts at the armhole and are thus not part of the sleeve seam.) Cast off in patt.

Left Sleeve
Work exactly as given for right sleeve, but work 56 rows of Chart E for the left sleeve, instead of chart for right sleeve.

Alternative Sleeves
If you would like sleeves without cow motifs work in the following way: using 3¾mm (9) needles and Cream Aran, cast on 47 sts. Work 28 rows in bobble rib as given for back, inc 16 sts evenly over last row, 63 sts.
Change to 4½mm (7) needles.
Next row K inc 17 sts evenly over row.
Next row P (80 sts).
Now work in cable and bobble patt as follows:
Row 1 (Right side) k1, *work row 1 of bobble patt over next 3 sts, work row 1 of cable patt over next 12 sts, rep from * to last 4 sts, work row 1 of bobble patt over next 3 sts, k1.
Row 2 P1, *work row 2 of bobble patt over next 3 sts, work row 2 of cable patt over next 12 sts, rep from * to last 4 sts, work row 2 of bobble patt over next 3 sts, p1.
These 2 rows set the patt. Now work in patt as established, working appropriate rows of cable and bobble patts and inc 1 st at each end of 3rd and every foll 4th row 15 times (110 sts) and then inc 1 st at each end of every alt row 15 times (140 sts). (Inc on 3rd stitch from each end by the Make 1 method.) Take the extra sts into the patt as they become available. Then continue straight in patt without further inc until sleeve measures 51cm (20in) or desired length from cast-on edge. Cast off in patt.
Join shoulder seams Put 37 sts from the back shoulder stitch holder and the same from the front shoulder stitch holder onto 4½mm (7) needles. Place these two needles side by side with the wrong sides of the work facing each other. Then working on the right side of the work, with another 4½mm (7) needle, k tog 1 st from each needle to give 1 st on the right-hand needle, *k tog the next 2 sts (now 2 sts on the right-hand needle), then pass the first of these sts over the second. Rep from * to work the rest of the sts. This creates a ridge on the right side of the work for the shoulder seam.

Neckband
Using 3¾mm (9) needles and Cream Aran with right side facing and starting at right front neck edge, k 15 sts from holder at centre front neck edge, pick up and k 14 sts to shoulder seam, k across 49 sts on holder for back neck, dec 4 sts evenly across these sts, pick up and k 14 sts down left side of neck, k across 15 sts on holder, 103 sts. P 1 row. Work 6cm (2½in) in bobble rib as folls:
Row 1 (Right side) *k3, p2, k3, p2, rep from * ending k3.
Rows 2, 4 and 6 *P3, k2, p3, k2, rep from * ending p3.
Row 3 *K1, MB in next st, k1, p2, k3, p2, rep from * ending k1, MB in next st, k1.
Rows 5 and 7 As row 1.
Row 8 As row 2.
When collar measures 6cm (2½in) cast off in patt.

To Make Up
Sew sleeves into armholes, the last 7cm (2¾in) at top of sleeve fitting into the cast off sts at armhole on back and front to make a neat right-angle. Join side and seam sleeves. Darn in any loose ends. Press seams. Sew on buttons. Embroider eyes on cows.

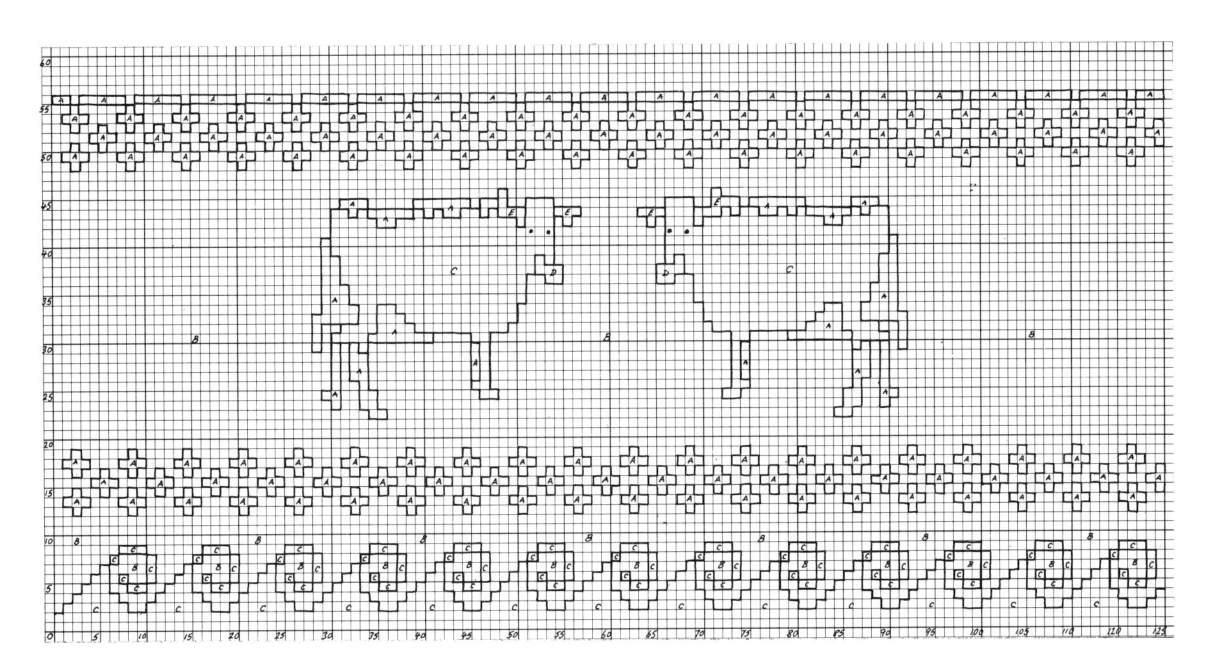

Chart A

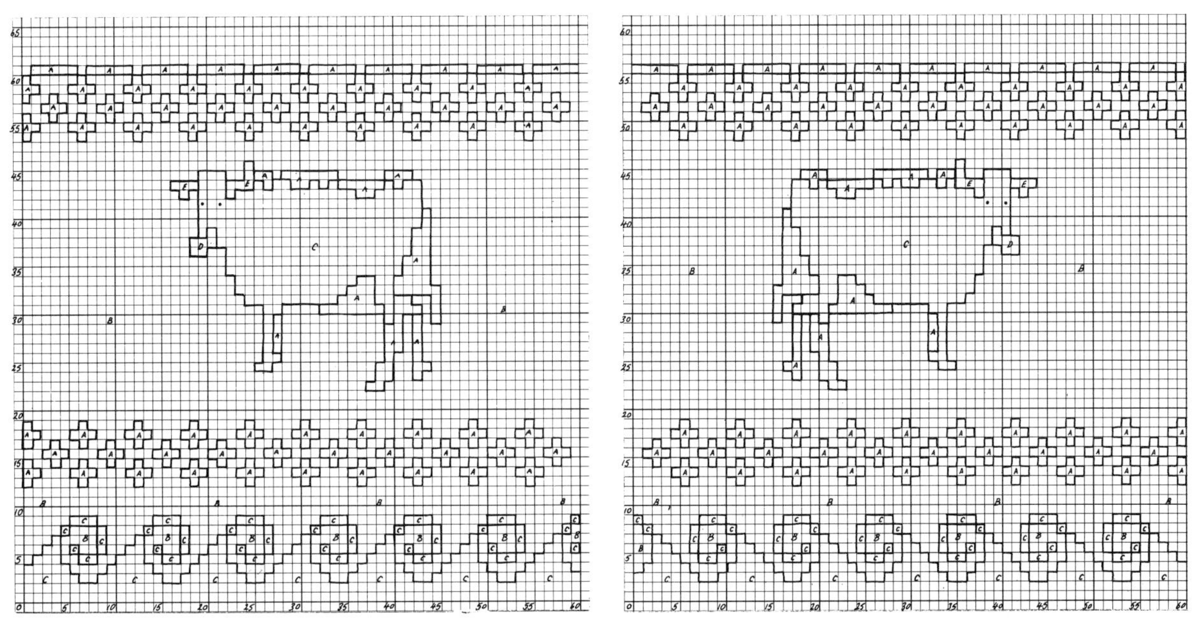

Chart B

Chart C

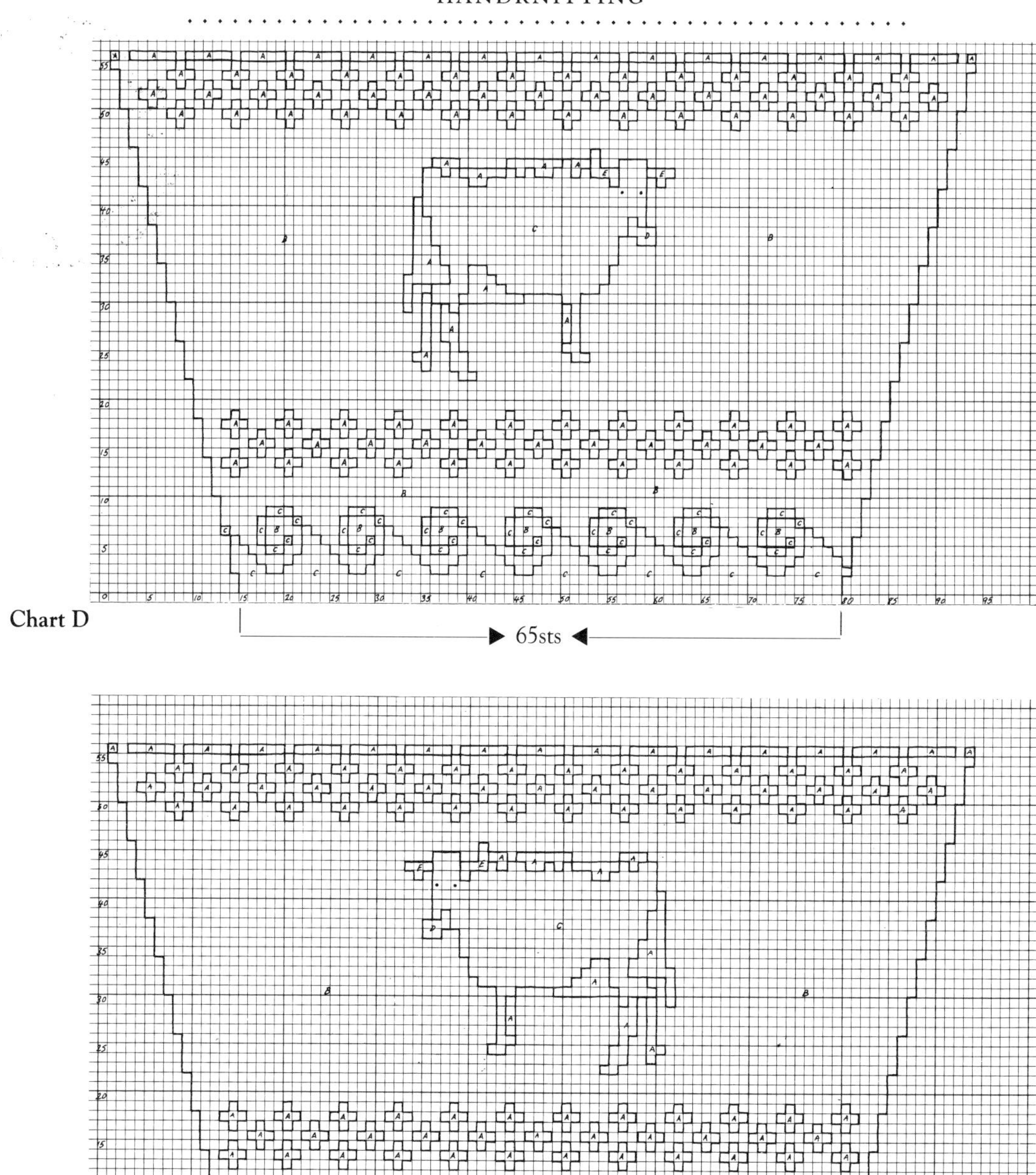

Chart D

Chart E

Mohair Triangle Coat

BY MELINDA COSS

This lovely bright mohair jacket is quick and easy to knit as long as you have basic intarsia skills. But it is also a good design to make if you want to try some simple intarsia for the first time.

MEASUREMENTS
One size only
Fits loosely up to bust 107cm (42in)
Actual measurement 137cm (54in)
Length 79cm (31in)
Sleeve approx 49cm ($19\frac{1}{4}$in)

MATERIALS
Melinda Coss Mohair yarn as follows:
Main Colour:
Royal 391g
Contrast Colours:
Red 109g
Yellow 42g
Cerise 23g
Purple 13g
Green 13g
Orange 13g
Turquoise 13g
One pair each of $4\frac{1}{2}$mm (7) and $5\frac{1}{2}$mm (5) needles
4 buttons

TENSION
17 sts and $18\frac{1}{2}$ rows for a 10cm (4in) square in st st on $5\frac{1}{2}$mm (5) needles.

ABBREVIATIONS
beg–begin(ning), **cont**–continue, **foll**–follow(ing), **inc**–increase, **k**–knit, **MC**–main colour, **patt**–pattern, **p**–purl, rem–remaining, **rep**–repeat, **st(s)**–stitch(es), **st st**–stocking stitch, **tog**–together, **yrn**–yarn round needle.

Back
With $4\frac{1}{2}$mm (7) needles and Royal, cast on 89 sts.
Rib row 1 (Wrong side) k1, *p1, k1, rep to end
Rib row 2 P1, *(k1, p1) to end
Rep last two rows for 12cm ($4\frac{3}{4}$in), ending rib row 2 and inc 23 sts evenly across last row (112 sts).
Change to $5\frac{1}{2}$mm (5) needles and beg patt.
1st row With red, k first 2 sts from Chart A, with MC k108, change to red and work 2 sts from Chart B.
2nd row P4 red, p104 MC, p4 red.
The sts are now set for Charts A and B. Work 18 rows of both charts. Then rep charts up back of jacket, next charts worked with yellow, rep with purple, green, orange, cerise and finally turquoise.
Shoulder shaping Cast off 20 sts at beg of the next 4 rows.
Cast off rem 32 sts.

Left Front
With $4\frac{1}{2}$mm (7) needles and MC cast on 45 sts.
Work in rib as for back and inc 11 sts evenly across last row (56 sts). Change to $5\frac{1}{2}$mm (5) needles and beg patt:
1st row With red, k first 2 sts from Chart A (p. 51), with MC 52 sts, change to red and work 2 sts from Chart B (p. 51).
2nd row P4 red, p48 MC, p4 red.
The sts are now set for Charts A and B. Work 18 rows of both charts. Then rep charts up left front, next 2 charts worked in yellow, followed by purple then work 4 rows of Charts A and B with green.
** **Shape Front** K10 green, k34 MC, k2 tog, k10 green.
Work three more rows of charts.
Next row K18 green, k17 MC, k2 tog, k18 green.
Cont working charts with the sequence as for back – at the same time dec 1 st after working contrast sts on next and every foll 4th row after

working neck edge sts, until there are 40 sts.
Work until 18th row of turquoise patt, ending on wrong side row.
Cast off 20 sts at beg on the next 2 rows.

Right Front
Work as for left front to **.
Shape front K10 green, with MC k2 tog, k34 MC, k10 green.
Work three rows of chart.
Next row K18 green, k2 tog, k17 MC, k18 green.
Cont working charts, cont working on right next edge after working contrast sts, until there are 40 sts.
Work until 17th row of turquoise patt, ending on a right side row.
Cast off 20 sts at beg of the next 2 rows.

Sleeves
With $4\frac{1}{2}$mm (7) needles and MC cast on 39 sts.
Work 10cm (4in) in rib as for back, ending rib row 2 and inc 17 sts evenly across last row (56 sts).
Change to $5\frac{1}{2}$mm (5) needles.
Work from Chart C in st st beg with a k row – at the same time inc 1 st at either end of the 3rd and every foll 4th row until there are 88 sts.
Continue straight until sleeve measures 49cm ($19\frac{1}{4}$in), ending on a wrong side row.
Cast off loosely. Join shoulder seams.

Button Band and Half Collar
With $4\frac{1}{2}$mm (7) needles and red, cast on 17 sts.
Work in rib as for back until band, slightly stretched, fits up left front to beg of neck shaping, ending wrong side row.
Shape for collar Next row: rib 2, then k1, p1 and k1 all into next st, rib to end (19 sts).
Rib 3 rows.
Rep last 4 rows 14 times (47 sts).
***Cont straight in rib until collar, slightly stretched, fits round to centre back neck.
Cast off loosely in rib. Sew band and collar in position.
Mark positions on band for 4 buttons, first 2cm(1in) above cast-on edge, last at beg of collar shaping and rem 2 spaced evenly between.

Buttonhole Band and Half Collar
Work as button band to beg of neck shaping making buttonholes correspond with markers as folls:
Buttonhole row (Right Side) rib 8, yrn, p2 tog, rib to end.
To shape collar – next row (Right side) rib to last 3 sts, then k1, p1 and k1 all into next st, rib 2. Rib 3 rows. Rep last 4 rows 14 times (47 sts).
Work from *** to end as for button band and half collar.
Join the centre back seam of the collar. Sew on sleeves, placing centre of sleeves to shoulder seams. Join side and sleeve seams. Sew on buttons.

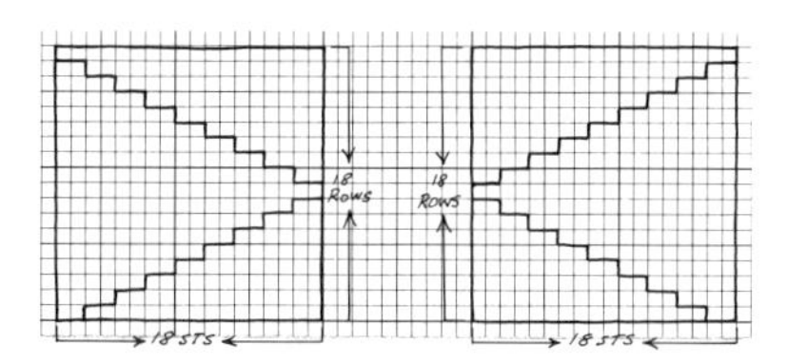

Chart A **Chart B**

Colour sequence up jacket: red, yellow, purple, green, orange, cerise and turquoise.

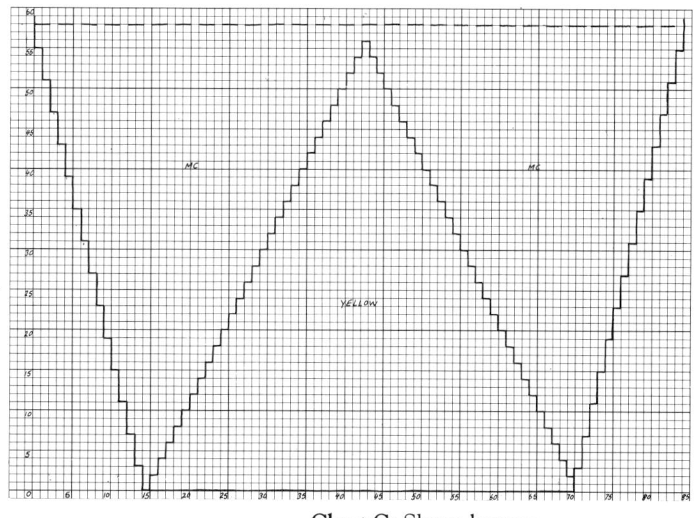

Chart C: Sleeve bottom

Bill Cardy for Men

BY MARION FOALE

This beautiful unisex sweater will appeal to many knitters. It is quick to knit and could be tackled by a relative newcomer as long as an even tension is maintained and as much attention is paid to making up the garment as to knitting it. The finishing off is very important and, if well done, will give a stylish professional finish. This pattern is suitable for Rowan Yarns quality Designer DK and can be adapted to fit women.

MEASUREMENTS

Chest size	*Widest point*	*Finished length*	*Sleeve length*
Small	117cm (46in)	57cm (27in)	53cm (21in)
Medium	123cm (48in)	58cm (27½in)	55cm (22in)
Large	128cm (50in)	59cm (28in)	57cm (22½in)
Extra Large	134cm (52in)	61cm (28½in)	59cm (23in)

For Ladies' Sizes
Use a size smaller as follows:
Small for medium
Medium for large
Large for extra large
Knit the sleeves 6cm (2½in) shorter than instructions by working 3 dec rows less, dec extra sts on needle to cuff sts.

MATERIALS

16(16,17,17) 50 gram balls DK
Pair each of 4mm (8) and 3¼mm (10) needles
Selection of spare stitch holders
6 buttons
Materials to make pocket lining.

TENSION

22 sts and 30 rows for a 10cm (4in) square in st st on size 4mm (8) needles.

ABBREVIATIONS

alt–alternate, **beg**–begin(ning), **cont**–continue, **dec**–decrease, **foll**–follow(ing), **inc**–increase, **k**–knit, **p**–purl, **rep**–repeat, **st(s)**–stitch(es), **st st**–stocking stitch, **tbl**–through back of loop, **tog**–together, **y.fwd**–yarn forward.

Back
With size 3¼mm (10) needles cast on 105(111,117,123) sts. Work 34 rows in k1, p1, twisted rib, (k, into back of k st). Change to size 4mm(8) needles and continue in st st inc 1 st at each end of the 5th and then every foll 4th row as follows:
(Right side) k2 inc into next st, k to last 4 sts, inc into next st, k3, until there are 127(133,139,145) sts. Work a further 45 rows.

Shape Armholes
(Right side) mark each end for start of sleeve. Now dec 1 st at each end of the next and then every foll alt row as follows:
(Right side) k2, k2 tog, tbl, k to last 4 sts, k2 tog, k2 until 111(117,123,129) sts rem.
Work straight a further 62(66,70,74) rows.

Shape shoulders
(Right side) p38 (41,44,47) sts, and leave on spare st holder for shoulder seam, cast off next 35 sts (back neck), p remaining 38(41,44,47) sts and leave on spare st holder.

Left Front
With size $3\frac{1}{4}$mm (10) needles cast on 48(51,54,57) sts and work in k1, p1 twisted rib as follows:
1st row (Right side) (Size medium and extra large p1) *k1, p1, rep from * to last 2 sts. K2.
2nd row *K1, p1, rep from * to last 2(1,2,1) sts, k2, (k1, k2, k1). Work 34 rows in all.
Change to size 4mm (8) needles and cont in st st. Inc 1 st at side edge on the 5th and every foll 4th row as follows:
(Right side) k2 inc into next st, k to end of row, *at the same time* when 22 rows have been worked (53,56,59,62 sts):
Work pocket opening (Right side) Work 13(16,19,22) sts. Leave these on a spare st holder and cont on remaining 40 sts, work to end of row.
1st row (Wrong side) P to last 2 sts, p2 tog.
2nd row K2 tog, k to end of row.
Rep these 2 rows until 15 sts remain (25 dec in all). Leave sts on a spare st holder and cont on remaining 13(16,19,22) sts.
Keeping continuity of side inc as set work as folls:
1st row (Wrong side) inc into first st, p to end of row.
2nd row K to end of row, inc into last st. Rep these 2 rows to work 25 inc in all 44(47,50,53) sts.
Next row (Right side) now working straight at side edge, k across 44(47,50,53) sts and then across 15 sts on stitch holder 59(62,65,68) sts.
Work straight for a further 41 rows.

Shape Armhole
(Right side) mark side edge of row for start of sleeve. Now dec 1 st at the beg of the next and then every foll alt row as follows:
(Right side) k2, k2 tog, tbl, k to end of row until 51(54,57,60) sts remain. Work a further 3 rows (all sizes).

Shape Neck
(Right side) now dec 1 st at neck edge on the next and then every alt row as follows:
(Right side) k to last 4 sts, k2 tog, k2 until 38(41,44,47) sts remain. Work a further 35(39,43,47) rows. Leave sts on a stitch holder for shoulder seam.

Right Front
Work the same as left *reversing shaping* and *pocket position.*
Working side inc as folls:
(Right side) k to last 4 sts, inc into next st, k3.
Working armhole dec as follows:
(Right side) k to last 4 sts, k2 tog, k2.
Working shape neck dec as follows:
(Right side) k2, k2 tog, tbl, k to end of row.

Shoulder Seam (Work 2 the same)
With size 4mm (8) needles put 38(41,44,47) sts from the back and the same from the front onto spare needles. Place these two needles side by side with the wrong sides of work facing each other. Then working on the right side of work k tog a st from each needle to give 1 st on right-hand needle. *K tog the next 2 sts (now 2 sts on right-hand needle) then pass the first of these 2 sts over the second. Rep from * to work the rest of the sts.

Sleeves (Work 2 the same)
(Right side) starting and ending on marked rows. With size 4mm (8) needles, pick up and k 109(115,121,127) sts evenly along armhole edge (centre st from shoulder seam), now work in st st dec 1 st at each end of every 5th row as folls:
On k rows K2, k2 tog, tbl, k to last 4 sts, k2 tog, k2.
On p rows P2, p2 tog, p1 to last 4 sts, p2 tog, tbl, p2 until there are 61(65,69,73) sts rem. Work 2(3,2,3) rows.
Shape to cuff (Wrong side) dec 10 (12,14,16) sts evenly across row (51,53,55,57 sts). Change to size 3¼mm(10) needles and work 34 rows in k1, p1, twisted rib.
Cast off in rib.

Pocket Top (Work 2 the same)
With size 3¼mm (10) needles, right side facing, pick up and k29 sts along pocket edge. Work 8 rows in k1, p1, twisted rib.
Cast off in rib.

Pocket Lining (Work 2 the same)
With size 4mm (8) needles, right side facing, pick up and k29 sts along pocket lining edge and work 50 rows in st st. Cast off.

Front Bands and Collar (all in one)
With size 3¼mm (10) needles cast on 15 sts. Work in k1, p1, twisted rib as folls:
1st row (Right side) *k1, p1, rep from * to last st, k1.
2nd row K1, *k1, p1, rep from * to last 2 sts, k2. Work until border measures 8cm (3in) less than front edge to first neck dec.
Mark button position The first on the 5th row, the second on the 27th row, then 4 more equally spaced, the last to fall on last side row.
Now shape collar edge Cont in rib as set inc 1 st at collar (outside) edge on the next and then every alt row as folls:
(Right side) (k1, p1) twice, k and inc into next st, rib as set to end of row, *at the same time* when 12 rows have been worked mark this row on the edge to be sewn to front. Work 28 inc in all (43 sts). Now work straight in rib as set for 113(121,129,137) rows.
Shape collar edge Cont in rib as set dec 1 st at neck edge on next and then every alt row as folls: (Right side) (k1, p1) twice, k2 tog, tbl, rib as set to end of row. Work 28 dec, in all (15 sts) *at the same time* mark corresponding row to match other side. Continue in rib as set, working buttonholes to correspond to button positions.
To work a buttonhole (Right side) work 7 sts, y.fwd, k2 tog, work to end of row. Cast off in rib.

To Make Up
Work in all ends. Fold pocket lining in half, sew sides, to make a bag, and top to bottom edge of pocket welt. Sew through sleeve seam and side seams. Attach front band and collar (marked row should be placed to first neck shaping). Ease front edge to stretched band from cast on to marked rows (collar should sit comfortably along remainder). Sew buttons to marked positions.

MACHINE KNITTING

The World's first knitting machine was invented in Britain in 1589 although it was not until a couple of centuries later that machines became popular in the garment industry. Now, machine knitters at home have some very advanced technology at their fingertips and can produce knitwear just as good as that made commercially.

Machine knitting is another very important part of this book and we have included several designers and technicians.

Children's designs are very popular and Sally-Anne Elliott has a beautiful guernsey for babies whilst Anna Davenport has some clever mouse designs.

The Fair Isle tradition has given inspiration to John Allen for his classic Fair Isle cardigan while Sophie Schellenberg's design shows the influence of the Yorkshire countryside in which she lives.

Both Hilary Highet and Ruth Herring have gone for a high fashion approach whilst Mary Davis uses sumptuous silks to create a stunning jacket. There is also an opportunity for those wanting to experiment with yarns and their machines, to create Vikki Haffenden's cushion which is a visual delight.

Classic Fair Isle Cardigan

BY JOHN ALLEN

This attractive classic button-through Fair Isle is simple and quick to make and is ideal for less experienced machine knitters, but you must have a machine with a punchcard and ribbing attachment.

MEASUREMENT

To fit 112cm (44in) chest
Sleeve (Centre body to cuff) 84cm (33in)
Length 60cm (24in)

MATERIALS

A	300g	Black wool = MY
B	150g	Deep pink chenille
C	125g	Peach chenille
D	125g	Beige chenille
E	100g	Green wool
F	100g	Navy wool

Waste yarn

NOTES

The punchcard must be made exactly as the chart (see p. 58) so that the sweater will match up at the seams. Shoulder seams are set back slightly to allow for grafting on the plain row and continuity of pattern.

TENSION

14 sts and 14 rows for a 5cm (2in) square over pattern.

ABBREVIATIONS

dec–decrease, **k**–knit, **MY**–main yarn, **Ns**–needles, **st(s)**–stitch(es), **WP**–working position, **WY**–waste yarn

Back

Using ribbing attachment cast on 169 sts. K 41 rows. Push 85 Ns into WP to left of 0 and 84 to right of 0. Set punchcard to 1, transfer sts from rib on to Ns, k 81 rows. Take 8 sts from each side of work off the machine on to WY. Take 8 Ns on each side out of work. Dec 1 st every 2 rows on each side, then k 57 rows to row 146. K off on WY. You should end knitting on 2 rows of MY only.

Left Front

Using ribbing attachment cast on 86 sts. K 41 rows. Push 86 Ns into WP, 49 left of 0 and 37 right. Set punchcard to row 1. Transfer sts from rib on to Ns and k 81 rows. Take 8 sts from left side of work off the machine on to WY and take 8 Ns out of work. Dec 1 st 4 times every 2 rows on left side, then k 43 rows to row 132. Cast off 22 sts on right side of work on to WY, take 22 Ns out of work. Dec 1 st every 2 rows 5 times on right of work, k 11 rows to row 153. K off on WY. Hand graft shoulder seam. (You should knit off on last row of small pattern.)

Right Front

Using ribbing attachment cast on 86 sts. K 41 rows. Push 86 Ns into WP, 37 left of 0 and 49 right of 0. Set punchcard to row 1, transfer sts from rib on to Ns, k 81 rows. Take 8 sts from right side of work off the machine on WY, take 8 Ns out of work. Dec 1 st, 4 times every two rows on right side of work, then k 43 rows, to row 132. Cast off 22 sts on left side of work on to WY. Take 22 Ns out of work. Dec 1 st every 2 rows 5 times on left side of work, then k 11 rows to row 153, k off on WY. Hand graft shoulder seams.

Sleeves

Pick up 8 sts from each side of work, k up 1 st from sleeve edge each row to 130 plus the 16 already picked up from each side of the work. Set

punchcard at row 11 and k 145 rows decreasing evenly down sleeve, leaving 86 sts. K off on WY.

Cuffs
Using ribbing attachment cast on 80 sts and k 35 rows.

Neck
Using ribbing attachment cast on 150 sts. K 35 rows.

Left Front Facing
Cast on 18 sts, k 237 rows, cast off.

Right Front Facing
Cast on 18 sts, k 6 rows, * make buttonhole, k 10 rows and rep from * 10 times. After last buttonhole k 6 rows. Cast off.

Finishing
Hand graft cuffs to sleeves decreasing evenly across work to lose extra sts.

Finish double neck
Using MY pick up 22 sts from each side of work, 26 sts from right side of neck, 60 from back of neck and 33 sts from left side. Hand graft neck rib to sts from neck. Fold rib double and stitch down to inside. Using double thread sew side seam and sleeve seam through centre of edge stitch of work. Finish threads by passing through seam and looping through back of work. This not only gets rid of loose ends, but also tightens up the seam to correct tension. Finish threads around sleeve, join neck edges by looping through back of work. Sew on facings using mattress stitch. This may take some adjusting – be prepared to stretch facing slightly to fit work (front edge). Finish loose threads by looping through back of work.

A Black wool = MY
B Deep pink chenille
C Peach chenille
D Beige chenille
E Green wool
F Navy wool

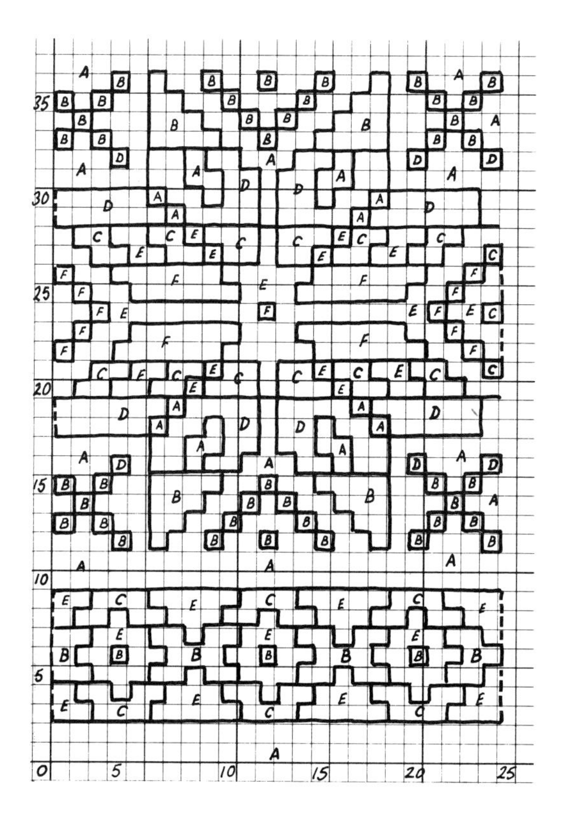

Botany Waistcoat

BY HILARY HIGHET

This stunning waistcoat is suitable for a machine knitter experienced in ribbed work who likes a challenge. It's worth taking time and care over the making-up to achieve a really professional finish.

MEASUREMENTS
Depth of rib 11cm (4in)
Length from centreback to bottom 52cm (21in)
Bottom back width 32cm (13in)

MATERIALS
Yeomans 4-ply Botany Sport (Ink) – 250g
Embroidery yarns
12 small buttons

TENSION
Body $32\frac{1}{2}$ stitches and 44 rows = 10cm (4in)
Welt 40 stitches and 48 rows = 10cm (4in)

ABBREVIATIONS
beg–begin(ning), **carr**–carriage, **COL**–carriage on left, **COR**–carriage on right, **dec**–decrease, **k**–knit, **H**–half pitch position of ribber bed, **HP**–holding position, **MB**–main bed, **MT**–main tension, **Ns**–needle stitches, **NWP**–non-working position, **0**–centre of bed, **P**–pitch position of ribber bed, **RC 000**–set row counter to 000 (or number given), **st(s)**–stitch(es), **T**–tension, **T0/0**–the first number refers to the tension on the main bed, and the second number to the tension on the ribber, **WP**–working position.

NOTE
Use close knit bar throughout.

Left Front
Refer to Fig. 1, p. 61.
Cast on and set up for 2 × 1 industrial rib 39. 0. 26, COR *but* push two end Ns on (on right of carr only) from NWP to WP. K zigzag row T2/2.
Hang comb and weights. Set machine for circular rows. K 2 circular rows. COL.
Alter carr to k rib T4/T4. H6 RC 000. Rib 53 rows.
Push all Ns up for full needle rib. Check that end Ns on ribber are correct – it may be necessary to make an adjustment here.
T4*/T4*. RC 000, k 1 row COL.
For the decorative panel, transfer sts from ribber to MB. 7 sts 0. Ensure ribber Ns are completely in NWP. K 1 row.
For the lace motifs, RC 002, COR. Beg transferring by hand as shown on Fig. 2, p. 61.
Continue this on every 2nd row until front is complete. Continue straight:
RC 024: Begin decreasing using 3-pronged transfer tool. Dec 1 st from both beds, on left side of bed only. Continue decreasing every 2nd row until 26 sts 0 (RC 048).
Continue on these Ns (26 st 0), remembering lace to RC 160.
For the shoulder, set both main and ribber carriages for HP COR. * Push 6 sts at opposite end to carr into HP. Knit 1 row (wind yarn around last Ns in HP to prevent hole). Repeat from * 6 times more. 10 sts remain.
* Push 5 sts at opposite end to carr into HP. K 1 row. Wind yarn around last Ns in HP to prevent hole. K 1 row.
To complete, set both carr to knit back all Ns in HP. K 1 row.
Transfer all sts from ribber to MB. MT + 4, knit 1 row.

Right Front
As left front *but* begin with COL 26 0 39. (For COR read COL, etc.)
Make buttonholes on rows 4, 20, 36 and 52.
To make buttonholes transfer A to B and C to D (see Fig. 3, p. 61). Knit through.

Keep empty Ns in WP. Knit 1 row.
Twist loop on empty N on MB only.
Continue for other buttonholes.

Back

Refer to Fig. 4, p. 61. 64sts 0. Set up for 2 × 1 rib. T0/0. Knit zig-zag row. T2/2 knit 2 circular rows. COL. Set machine for rib T4/4. Rib 53 rows. For the main body, RC 000. Push all Ns up for full needle rib. T4*/T4*. Continue straight:
RC 024. Begin decreasing using 3-pronged transfer tool, dec 1 st from each bed at both ends of work. K 2 rows. Repeat 11 more times; RC 048. 52sts 0 left in works.
Continue straight to RC 120.
To make darts, set both carrs for HP. Push 8 Ns at opposite end to carr into HP at beg of next 6 rows. Remember to wind wool around last N in HP to prevent holes. 24 sts at each side of work now in HP.
Return 24 sts at opposite side to carr to WP. K 1 row. Set carr to k back Ns in HP. K 1 row. RC 128. All Ns back in work.
To complete back, k straight to RC 160. Set for HP. Push 4 sts at opposite end to carr into HP at beg of next 18 rows. (16 st 0 left in work.)
Return Ns to work after next 2 rows (as for darts). Transfer all sts to MB. Knit 1 row MT + 4. Cast off with latchet tool.

To Make Up

Steam and block all pieces.
The front shoulder is pleated to fit the back shoulder and backstitched into position.
For the side seams, either mattress stitch or backstitch carefully in 'channel' of knitting.
Embroider 'lace' as you choose (see photo, above and Fig. 5, p. 61). Sew on buttons. Finish ends and press.

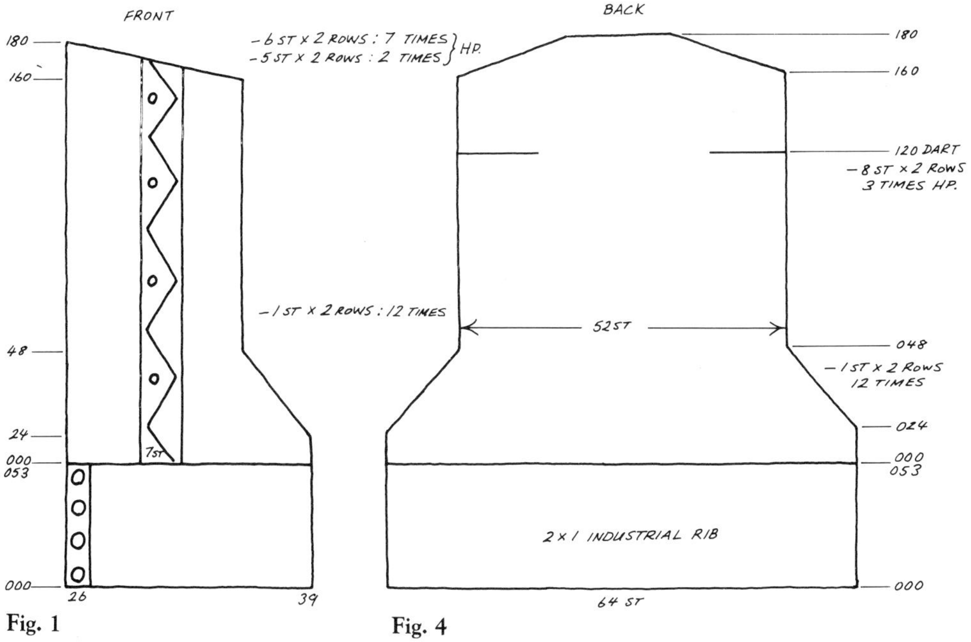

Fig. 1

Fig. 4

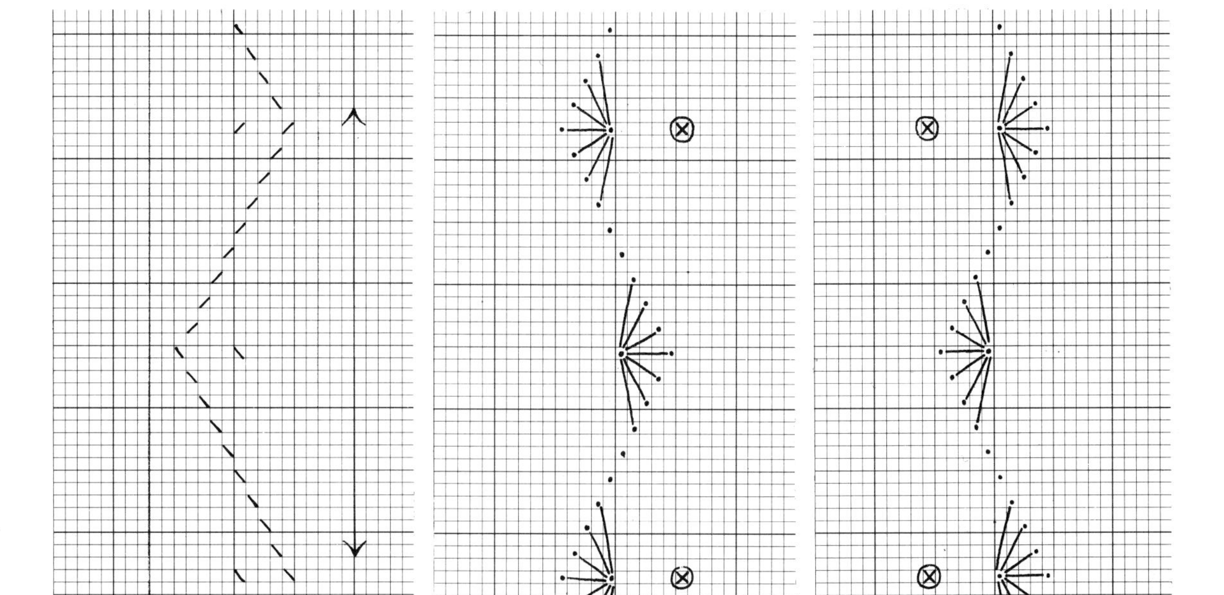

Fig. 2

Fig. 5

C B

D A

Fig. 3

BUTTONHOLES Rows 4, 20, 36, 52

Lamberan

BY RUTH HERRING

This challenging, brightly coloured waistcoat features an intarsia design on the front inspired by rug designs from the Lamberan region of Persia. The back is made in a simple punchcard design.

MEASUREMENTS

To fit 80–95cm (32–38in) chest
Total length 52cm (21in)

MATERIALS

100% Cotton Designer 4-ply yarn (Yorkshire Mohair Mills) as follows:

A	100g	Black = (MY)	22
B	100g	Red = (CY)	17
C	50g	Steel	11
D	30g	White	00
E	30g	Khaki	07
F	30g	Peach	24
G	30g	Green	15

Waste yarn in a contrast colour (WY)

TENSION

T6 – 33 sts and 37 rows to 10cm (4in) over pattern. Check tension carefully.

ABBREVIATIONS

beg–begin(ning), **carr**–carriage, **cont**–continue, **CY**–contrast yarn, **dec**–decrease, **foll**–follow(ing), **HP**–holding position, **inc**–increase, **k**–knit, **MY**–main yarn, **N**–needle, **patt**–pattern, **pos**–position, **rep**–repeat, **RC 000**–set row counter to 000 (or number given), **T**–Tension, **WY**–waste yarn.

NOTES

The punched pattern used was Brother standard design No 2j. Any simple punchcard pattern can be substituted. When working intarsia pattern, wind off small amounts of each colour. When joining in a new colour leave an end of approx 5cm (2in) for darning in later.

Back

Using WY, cast on 149 sts and work a few rows.
Change to MY, work 6 rows. T5.
Work 1 row T10.
Work 6 rows T5 (locking in punchcard on last row). Pick up row 1 MY to turn hem. Remove WY and hang on weights.
RC 000: Change to T6, with MY in feeder 1/A and CY in feeder 2/B, k punchcard patt inc 1 st each side of rows 11, 21, 31, 41 (157 sts). Cont straight to RC 62.
Shape armholes Cast off 3 sts at beg of next 2 rows, then dec 1 st each side of next and every row to RC 83. Then dec 1 st each side of every 4th row to RC 107. Cont straight to RC 166.
Shape shoulders: Set carr to HP *. Push 4 needles on opposite side to carr to pos E (HP) Take carr across once and slip yarn under first held N to avoid making a hole. Rep from * 8 times. Leave 29 sts for each shoulder and 53 sts for back neck on WY.

LEFT FRONT

RC 000 * Using Steel, cast on 3 sts using closed edge method. Work 1 row T6. Inc 1 st on left side of work, work 1 row. Rep from * once.
Set carr to intarsia and lay all yarns at feet in front of machine. Using chart (see pp. 64–5) as a guide and following solid guideline, work from row 5, casting on sts for front edge as set and laying in colours as indicated by symbols.
Shape side seam, armhole and shoulder as given for back.

RIGHT FRONT

RC 000, using Steel, cast on 3 sts using closed

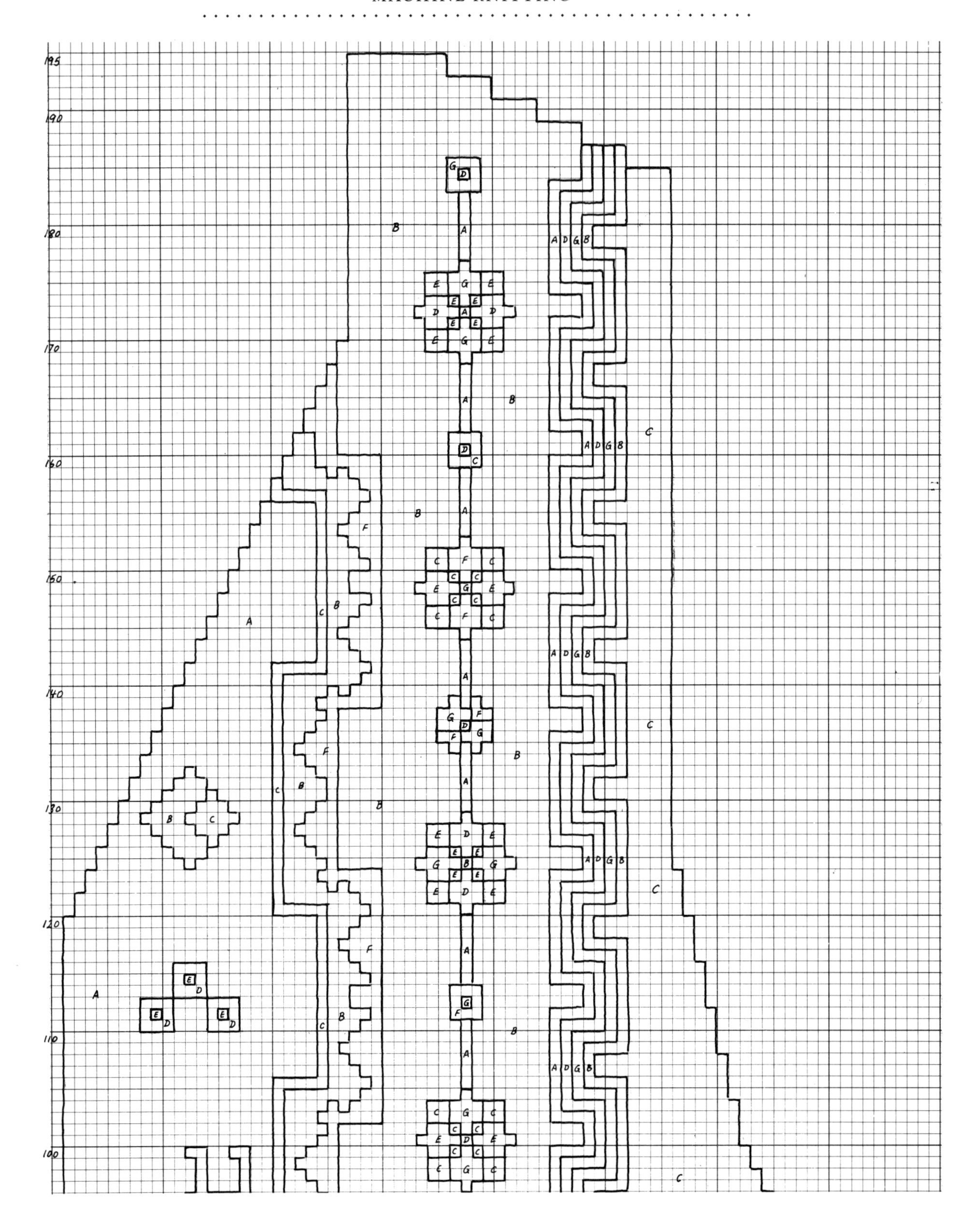
195
190
180
170
160
150
140
130
120
110
100
A
B
C
D
E
F
G
A D G B

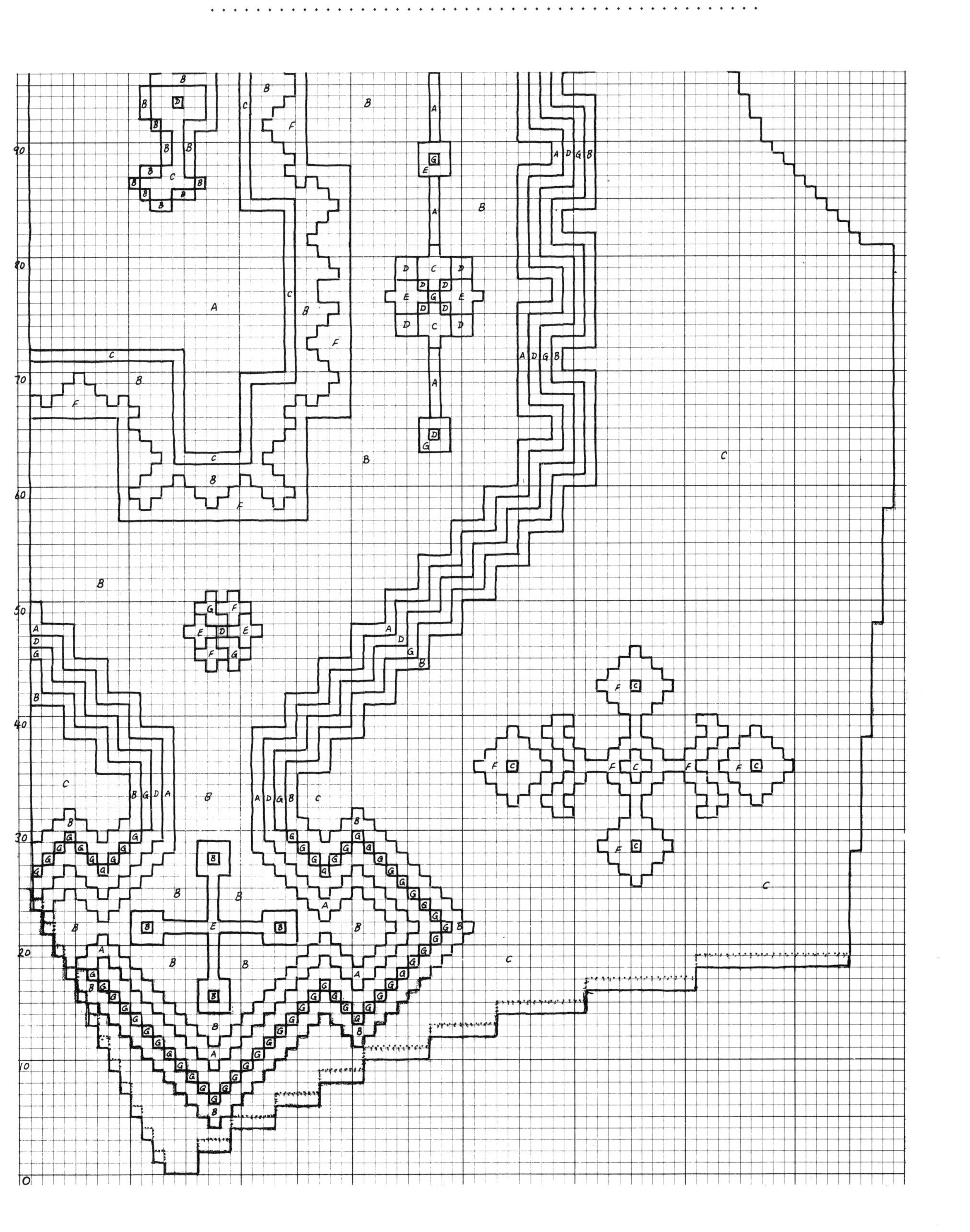
90
80
70
60
50
40
30
20
10
0

edge method, work 2 rows T6.
Inc 1 st on right side of work, work 1 row. Set carr to intarsia and work as given for left front from row 5 from chart (see pp. 64–5) reversing workings for left front chart and following dotted guideline. *Darn off ends.*
Left armhole band Join shoulder seam by picking up front and back sts for left shoulder. Work 1 row T5, cast off. With wrong side facing, pick up 82 sts evenly along armhole edge from shoulder seam to back side seam.
Rep for front armhole edge (164 sts).
MY, work 6 rows T5, 1 row T10, 6 rows T5. Pick up row 1 to form a double band. Work 1 row T5, cast off.
Rep for right armhole band.

Left Front Hems
Join side seams.
Long edge With wrong side facing and starting at side seam, pick up 60 sts to point at left front. Using MY, work 2 rows T5. * Inc 1 st at front point only on next row, work 1 row. Rep from * once.
Work 1 row T10.
** Work 1 row T5, dec 1 st at front point only in foll row. Rep from ** once. Work 2 rows. Pick up first row to form a double hem. Work 1 row T5, cast off.

Short edge With wrong side facing, pick up 19 sts from centre front edge to point. Work 2 rows T5. * Dec 1 st at centre front edge, inc 1 st at point. Rep from * until 4 shaping rows have been completed.
Work 1 row T10. ** Change to T5, dec 1 st at point, inc 1 st at centre front edge, rep from ** until 4 shaping rows have been completed. Work 2 rows. Pick up first row to form a double hem. Work 1 row T5, cast off.

Right Front Hems
Work as given for left front hems. Join hems at side seams.

Neck Band
Back neck and left front With wrong side facing and starting at right shoulder, pick up 3 sts down right back neck, 53 sts from WY, 3 sts up left back neck and 118 sts down left front, finishing at bottom of hem. Using MY, work 8 rows T5, 1 row T10, 8 rows T5. Pick up first row to form a double band. Work 1 row T5, cast off.
Right front With wrong side of right front facing, pick up 119 sts, from right shoulder to bottom of right front hem. Complete as given for left front and back neckband.
Join neckband at right shoulder and stitch double edges at front hems to neaten.

Machine Knitted Jacket with Silk Fabric Woven In

BY MARY DAVIS

. . . .

This beautiful wool jacket in Colinette One Zero with silk fabric woven in combines straightforward machine knitting with weaving and crochet to create a visually stunning garment.

. . . .

MEASUREMENTS

Width across chest 58 (66)cm/23 (26)in
Finished length 58 (66)cm/23 (26)in
Sleeve underarm 48 (53)cm/19 (21)in
N.B. All measurements are approximate. When knitted up, this yarn is very flexible.

MATERIALS

6–7(8–9) skeins of Colinette One Zero, Parrot Blue.
Approximately 1m (1 yard) each of 5 different colours of fabric. Silk may be dyed to match colours in yarn, but any oddments of washable fabric will do.

TENSION

T10 stocking stitch and needle arrangement as shown:
K on every other needle with 8th and 9th out of work, i.e.:
I.I.I.I..I.I.I.I..I.I.I.I..I.I.I.I..
for 20 rows. Put all empty Ns back into work for next 2 rows except left-hand N of each pair. When 2 rows are complete, replace remaining Ns in work and k 2 more rows, then transfer sts to restore N arrangement, as above. K 20 rows.
T10: 20 sts – 10cm (4in). 20 R – 18cm (7in).

ABBREVIATIONS

carr–carriage, **cont**–continue, **dc**–double crochet, **HP**–holding position, **inc**–increase, **k**–knit, **MY**–Main Yarn, **N**–Needle, **st(s)**–stitch(es), **T**–Tension, **tog**–together, **WY**–Waste Yarn.
Read instructions through carefully before starting to knit.

Back
Large size knitted in 2 pieces, exactly alike, but neck shaping reversed. See Fig. 1 p. 69.
Using 'E' wrap, closed edge method of cast on, cast on over full width of N bed, i.e. 114 sts (small size), 64 sts large size.
K in pattern as above to row 80(95).
Small size Shape back neck as follows: carr at right bring all Ns to left centre forward to HP, plus 16 to right centre. Set carr for Hold and K 4 rows.
Large size 18 Ns at left/right edge work to HP. Carr to Hold and K 4 rows. Remove right shoulder from machine as follows: break MY, thread WY and k 6 rows. Break WY and k 1 row across.
Small size Repeat from row 80 for left shoulder shaping, reversing instructions. This leaves centre neck sts on machine. Cast off firmly using spare end of yarn.

Front
K 2 alike, reversing neck shaping instructions. Cast on 56(64) sts using closed edge method as before and k as for back to row 70(85).
Shape front neck Carr at right cast off 10 sts at left edge of work. K 2 rows and cast off 2 sts. K 2 rows, cast off 1 st and continue in this way until 16(18) sts have been cast off. K to row 84(99) and remove shoulder from machine on 6 rows WY as before.
Shoulder seams Replace sts of left shoulder back on machine with right side work facing. Replace sts of left front shoulder on same Ns, right sides together. K 1 row T10 and cast off.

Sleeve
K 2 alike. See Fig. 2 below.
Cast on 70(80) sts as before, using closed edge method and arranging Ns according to pattern. K to row 72(82), inc 1 st at carr side of work every row until 100(110) sts, being careful to maintain N arrangement as diagram, throughout. At row 72(82) cast off loosely.

Making Up
Cut or tear fabric into strips approximately 4–5cm ($1\frac{1}{2}$–2in) wide, depending on thickness of fabric. Thread these strips, using a large safety pin as a needle, in and out of the ladders that have been formed lengthwise in the work by leaving 2 Ns out of work. When all the lengthwise ladders are complete, use 2 ends for one ladder and tie in a loose bow on surface of knitting, at random. Repeat weaving process for horizontal rows of holes. When weaving is complete, long ends of fabric on jacket body may be knotted or plaited. Beads may also be threaded on to the ends which hang down. The long ends of fabric on the sleeves may be twisted into one another and tied down or knotted close to the knitted fabric. Alternatively, the sleeve ends and hem of jacket may be hand-crocheted, incorporating the loose ends of fabric at the same time.

(Large size) Stitch centre back seam. (All sizes) Stitch sleeve shoulder edge to shoulder edge of body matching centre to shoulder seam and using a neat backstitch. Stitch through ends of fabric. Although this may offend purists, it is perfectly possible to make these seams using a domestic sewing machine with matching polyester thread and the longest straight stitch setting. Stitch side seam and sleeve underarm using same stitch. It is advisable to pin these seams first, putting the pins in at right-angles to the seam. Weave in all ends of MY. Stitch down loose ends fabric.

Neck edge
Replace all neck sts on machine 52(60), with wrong side of work facing and using all Ns. T10 k 20 rows. Cast off firmly. This is then allowed to roll.

Front Bands
Using MY crochet 2 rows dc. On right front band, next row, make loops for buttons by making a chain over 2 sts with 4 dc in between each buttonhole (8(9) buttonholes). Next row cont in dc. Each band 6 rows dc. Shoulder seams and neck edge may also be given more tension by 1 or 2 rows of dc if desired.

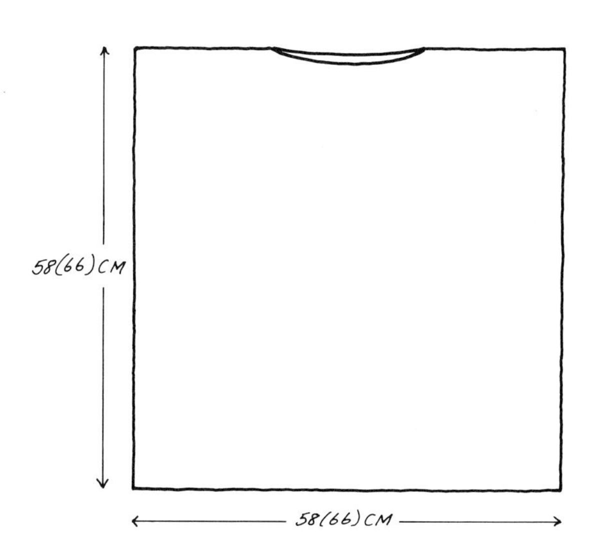

Fig. 1

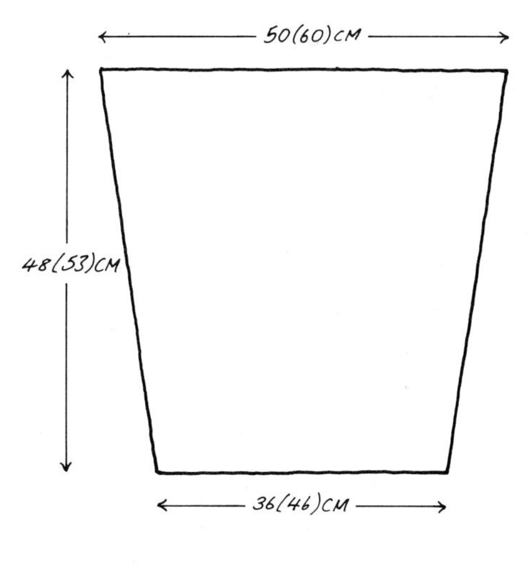

Fig. 2

Baby Guernsey

BY SALLY-ANNE ELLIOTT

This delightful baby guernsey makes an ideal project for an experienced machine knitter who is able to cable and do garter stitch. There is also a hat to accompany the jumper and make an attractive outfit.

MEASUREMENTS

To fit 6–9 months
Chest 61cm (24in)
Length 31cm ($12\frac{1}{4}$in)
Sleeve length 25cm (10in)

MATERIALS

One cone Rowan cable cotton

TENSION

31 sts and 36 rows for a 10cm (4in) square, set approximately at T7.

ABBREVIATIONS

alt–alternate, **dec**–decrease, **inc**–increase, **k**–knit, **MT**–main tension, **MY**–main yarn, **RC 000**–set row counter to 000 (or number given), **st(s)**–stitch(es), **st st**–stocking stitch, **WY**–waste yarn

CABLE NOTE

1 × 1 Cable; *cross one stitch over the adjacent stitch, k 2 rows*. Repeat from * to *.
Shaped cable:
Cable 1 × 1 k 2 rows, 3 times
Cable 2 × 2 k 4 rows, twice
Cable 3 × 3 k 6 rows, twice
Cable 2 × 2 k 4 rows, twice
Cable 1 × 1 k 2 rows, 3 times.

GARTER STITCH NOTE

These rows may be worked either by taking the knitting on to WY and replacing the stitches on to the same needles but with the reverse side facing, or with the use of a garter bar or a garter carriage if available or by transferring the stitches on to the ribber, knitting 1 row and transferring back to the main bed again.

Back

Push up 94 sts and cast on using WY, k a few rows then join in MY and MT – 2, k 10 st st rows, k 1 row garter st and then 10 rows st st. Pick up loops from first row in MY to form hem.
Change to MT and placing a 1 × 1 cable at the centre and 2 sts in from each edge, k to RC 74. RC 000, k 1 row garter st.
*K 1 row st st, place 1 × 1 cables over centre two sts and on sts 11 and 12, 16 and 17, and 28 and 29 each side of centre. Dec 1 st each end of every alt row 4 sts in from edge 7 times, then place 1 × 1 cable 2 sts in from edge each side. K to RC 46, k 6 rows garter stitch. Change to MT – 2 k 8 rows st st, one row garter st then 8 rows st st. Pick up loops from last row of garter st, k 1 row MT. Cast off taking the sts around gate pegs.

Front

K as for back until *. Place cables as follows, 1 × 1 over sts 11, 12, 16 and 17, each side centre, k 2 rows. Then commence shaped cable over centre 2 sts and sts 28, 29 each centre side. Continue as for back.

Sleeves

Push up 50 sts and k hem as for back. Place 1 × 1 cable over centre two sts. Inc 1 st each end of every 4th row 4 sts in from edge until 84 sts but every alt row, move 4 sts out, pick up loop from next st and dec 1 st at edge. K to RC 76. K 6 rows garter st. Place 1 × 1 cable over centre 2 sts. Dec 1 st every alt row 4 sts in from edge 7 times, cast off.

Making Up

Join armhole seams by picking up large loops from edge of knitting with wrong sides together (knit 1 row MT) and casting off both sets of sts together; do not include hems at neck edge. Join underarm and sleeve seams in the same way. Overlap neck welts to form envelope neck, stitch along neck starting at shoulder for 2cm ($\frac{3}{4}$in).

Hat

Cast on 60 sts using E-wrap method and MT, k to RC 40. K 1 row garter st (k 1 row) then place cables as follows, 1 × 1 cable 2 sts in from each edge, and shaped cable in the centre. K RC 84 then reverse shaped cable, RC 126. K 1 row garter stitch, k st st to RC 168, cast off. Join side seams by picking up loops from edge of knitting with wrong sides together and casting off both sets of sts together.

The Happy Couple

BY ANNA DAVENPORT

These delightful mice are very easy to knit on any standard gauge machine. It's the essential, careful making-up and finishing that will take the most time!

MEASUREMENTS

Bride and Groom 12cm ($4\frac{3}{4}$in) tall.
Bridesmaid 8.5cm ($3\frac{1}{4}$in) tall.

MATERIALS

Oddments of 4-ply acrylic yarn in white (W), black (BK), deep pink (P) and mauve (M)
Polyester fibre filling
Black and blue felt
Glue
1 metre (3ft) of 3mm ($\frac{1}{8}$in) wide white ribbon
Tiny bunches of artificial flowers
2.5cm (1in) circle of card

MAIN TENSION

Tension Dial approx 6.

ABBREVIATIONS

adj–adjacent, **alt**–alternate, **carr**–carriage, **dec**–decrease, **HP**–holding position, **k**–knit, **MT**–main tension, **N**–needle, **NWP**–non working position, **RC 000**–set row counter to 000 (or number given), **st(s)**–stitch(es), **st st**–stocking stitch, **WP**–working position, **WY**–waste yarn

NOTE

The bridesmaids may be knitted in any contrasting colour. Follow the first set of figures for the bridesmaid.

BRIDESMAID (BRIDE)

Dress, Beginning With Skirt

Push 38(50) Ns to WP. Using WY, cast on and k a few rows. RC 000. Using M(W) and MT – 1, k 18(24) rows.

Make shell edging * Push every 5th(6th) N to HP. Set carr to HP. Using MT + 2, k 3 rows. Set carr for st st. Using MT, k 1 row.

Transfer every 5th(6th) st on to adj N at right (the same N as was previously in HP) and push empty Ns to NWP. * k 18(24) rows. Using WY, k a few rows and release from machine.

Bodice

** Push 19(25) Ns to WP. With wrong side facing, rehang sts from last row worked in M(W), placing 2 sts on to each N (counting loops from Ns in NWP as a st). RC 000. ** Using M(W) and MT, k 7(10) rows.

Make Picot Neck Frill

*** Using MT + 2, k 2 rows. Transfer every alt st onto adj N at right. Leave empty Ns in WP, k 2 rows. Transfer every alt st onto adj N at right. Return empty Ns to NWP. Pick up every alt st from last row worked in MT and hang on to corresponding Ns. Using MT, k 1 row. Break yarn. Thread end through sts, remove from machine. ***

Head

Push 16(22) Ns to WP. Using WY, cast on and k a few rows. RC 000. Using P and MT, k 10(14) rows. Using two-pronged transfer tool, dec 2 sts both ends of next row. Using two-pronged transfer tool, dec 1 st both ends of next and every foll row, 4(6) times in all, RC shows 15(21), 4(6) sts. Break yarn. Thread end through sts and remove from machine.

Arm

(Knit two)

Push 9(12) Ns to WP. Using WY, cast on and k a few rows. RC 000. Using M(W) and MT, k 10 rows.

Bridesmaid Only
Using two-pronged transfer tool, dec 2 sts both ends of next row (5 sts). Using M, and MT, k 2 rows.

Bride Only
Using WY, k a few rows, release from machine. Push 6 Ns to WP. With wrong side facing, rehang sts from last row worked in W, placing 2 sts on to each N. RC 000. Using W and MT – 2, k 2 rows.

Bridesmaid (Bride)
Set machine for cord knitting. Using P and MT – 3, k 12(12) rows. Break yarn. Thread end through sts and remove from machine.

Veil (Bride Only)
Cast on 50 sts in W using the E-wrap method. RC 000. Using MT + 2, k 2 rows. Make shell edging as given for skirt from * to *. Using MT, k 10 rows. Using MT – 1, k 10 rows. Using MT – 2, k 30 rows. Using WY, k a few rows and release from machine.
Rehang as given for bodice from ** to **. Using MT, k 2 rows. Work picot frill as given for bodice frill from *** to ***.

Tail Bridesmaid (Bride)
Using P, cast on 2(2) sts using the E-wrap method. Set machine for cord knitting. Using MT – 2, k a length of about 10(14)cm/4($5\frac{1}{2}$in). Break yarn. Thread end through sts and remove from machine.

Ear
(Knit two)
Push 6(8) Ns to WP. Using WY only, cast on and k a few rows. RC 000. Using P and MT, k 4(4) rows. Pick up sts from first row worked in P and hang on to corresponding Ns, k 4(4) rows. Break yarn. Thread end through sts and remove from machine.

To Make Up
Remove WY from all pieces and thread a length of yarn through open loops. With right sides together, fold dress in half. Pull up sts at picot frill neckline and secure. Join centre back seam. Turn right side out. Sew a line of gathering sts along the waistline (row shown by dec sts). Place a little stuffing in bodice. Pull up waistline sts and secure. Fold skirt to the inside along shell edging, pull up remaining sts and secure to inside of waistline.
With right sides together, fold head in half. Pull up sts at tip of nose and secure. Join underside seam. Turn right side out and insert filling. Pull up remaining sts at back of head and secure. Attach head to body.
Pull up st in P at tip of hand and secure. With right sides together, join underside seam of sleeve in M(W). Place a little stuffing to pad out sleeve. Pull up sts at top of sleeve and secure. Make up remaining arm in the same way. Attach arms to body.
Pull up sts at base of ear and secure. Thread yarn from cast on edge of ear along one side, through gathered sts, along remaining side and pull up to

shape ear. Secure. Attach ears to head.
Attach tail to inside back of waistline. Tie ribbon around waist, bow and secure. Cut tiny circles for eyes and nose from felt and stick or sew in position.
Bride Only Block veil, pulling out points at lower edge of veil to shape. Steam lightly. Leave to dry out completely before removing pins. Pull up sts at top of veil and secure. Attach veil to top of head, positioning towards the back behind the ears. Using a length of white yarn, sew ends of hands together with a tiny bouquet of flowers secured between, and finish all bouquets with a bow of white ribbon.

BRIDEGROOM

Legs

First leg Push 16 Ns to WP. Using WY, cast on and k a few rows. RC 000. Using BK and MT, k 22 rows. Using WY, k a few rows and release from machine.
Second leg Push 16 Ns to WP at right of centre '0'. Work second leg as given for first but do not remove on WY.
Push 16 Ns at left of centre '0' to WP. With wrong side facing, rehang sts from first leg on to empty Ns. RC 000.

Body Coat, Beg With Coat Tail

K 2 rows. Set carriage to HP. Push 9 Ns at left and right side to HP. Cont over 14 Ns in centre for tail. K 8 rows. Dec 1 st both ends of next and foll 2 alt rows. K 1 row. Inc 1 st both ends of next and foll 2 alt rows. K 1 row. RC shows 22, k 8 rows. Set carr for st st. K 20 rows. Transfer every alt st onto adj N at right. Return empty Ns to NWP. K 1 row. Break yarn. Thread end through sts and remove from machine.

Arm

(Knit two)
Push 12 Ns to WP. Using WY, cast on and k a few rows. RC 000. Using BK and MT, k 12 rows. Using W and MT – 2, k 4 rows. Break yarn. Thread end through sts and remove from machine.

Head and Ears

Work as for bride.

Top Hat

Push 20 Ns to WP. Using WY only, cast on and k a few rows. RC 000. Using BK and MT + 1, k 2 rows. Using MT + 2, k 2 rows. Using MT + 3, k 1 row. Using MT + 2, k 2 rows. Using MT + 1, k 2 rows.
Make brim Pick up loops from first row worked in BK and hang on to corresponding Ns. Using MT – 1, k 10 rows. Transfer every alt st onto adj N at right. Return empty Ns to NWP. K 2 rows. Break yarn. Thread end through sts and remove from machine.

Tail

Work as for Bride to a length of 10cm (4in).

To Make Up

Remove WY from all pieces and thread a length of yarn through open loops. With right sides together, fold coat tail with side edges together and neatly join side seams. Turn coat tail right side out. Pull up sts at toes and secure. Join inside leg seams. Join centre front seam of body. Turn right side out. Stuff body and legs; do not place any stuffing in coat tail. Pull up sts at neck and secure. Fold ends of legs upwards for approx 1cm ($\frac{1}{2}$in) to create foot and mattress st in position. With right sides together fold head in half. Pull up sts at tip of nose and secure. Join underside seam. Turn right side out and insert filling.
Pull up remaining sts at back of head and secure. Attach head to body. With right sides together, fold arms in half. Pull up sts in W and secure. Join underside seam. Turn right side out and insert filling. Attach arms to body. Pull up sts at base of ear and secure. Thread yarn from cast on edge along one side, through gathered sts, along remaining side and pull up to shape. Secure. Attach ears to head.
With right sides together, fold hat in half. Pull up sts at top of hat and secure. Join centre back seam. Turn right side out. Position card circle in top of hat and stuff hat. Position hat on head and attach round inside edge of brim. Attach tail to back of body under coat tail. Tie ribbon around neck in a bow and secure. Cut tiny circles from felt for eyes and nose and stick or sew in position.

Machine Knitted and Felted Cushion Cover

BY VIKKI HAFFENDEN

This unusual cushion achieves its effect by felting the machine knitted fabrics. It is quick and easy to knit but time and care must be spent on machine-sewing the cushion edges.

This treatment of wool is the opposite of everything traditionally taught about the care of knitted fabric. The resulting fabric is firm and textural whilst the colours blend in together after washing, giving a soft, subtle glow to the finished article. The fabric is washable, and similar designs can be used to cover footstools and cushions that regularly get heavy wear from children, cats and general family use. Suitable for any 24 stitch standard gauge punchcard knitting machine (instructions are for Brother machines with those for Knitmaster/Silver Reed in brackets).

MEASUREMENTS

Finished size 38cm × 38cm (15½in × 15½in).

MATERIALS

2/8s Shetland Wool oiled on cone, approximately 100g in total (depends on your tension and felting technique).This is an ideal way of using up odd amounts of yarn after knitting garments. Adjust the colour changing to accommodate fewer colours if necessary.
Shetland Wool felts the best; however, if you have any pure wool oddments, it is worth trying a test square to see if it will felt or not.
Colour 1 = Kilncroft 60g
Colour 2 = Mustard 10g
Colour 3 = Regal Red 10g
Colour 4 = Petrol Blue 10g
Colour 5 = Apple 10g.
37cm (14in) zip in same colour as background
Sewing thread in same colour
38cm (15in) square cushion pad (feather or fibre)
(Punchcard graph A)

TENSION

After finishing: 31 sts = 10cm (4in), 32 rows = 10cm (4in).

ABBREVIATIONS

col–colour, **COL**–carriage on left, **COR**–carriage on right, **k**–knit, **KC**–knit card, **MY**–main yarn, **RC 000**–set row counter to 000, **st(s)**–stitch(es), **WY**–waste yarn.

Method

COR cast on 122 sts in WY, using weaving method or cast on comb (a closed edge is not necessary). K 20 rows in WY.
Thread col 1 (MY) in feeder A and k 3 rows plain.
COL outside the turn mark. Insert punchcard, lock on row 1, set dial to KC and k one row to select the needles for the first pattern row (see chart on p. 76). (Silver Reed owners – omit KC dial instruction, otherwise the same.)
Keeping col 1 (MY) in feeder A throughout, proceed as follows: COR, RC 000. Thread col 2 in feeder B, unlock card, push in MC buttons, knit 8 rows.**
(Silver Reed – omit MC instruction, side levers back, dial to F, otherwise the same.)
Colour 3 in feeder B – k 10 rows.
Colour 4 – 10 rows.
Colour 5 – 10 rows.
This is one colour repeat, you have used all the colours 2–5 in feeder B *once*. To make the pattern stagger, repeat the colour *order* but alter the number of rows knitted in each one as follows:
Colour 2 – 2 rows
Colour 3 – 8 rows

Colour 4 – 12 rows
Colour 5 – 8 rows
Colour 2 – 10 rows
Colour 3 – 10 rows
Colour 4 – 10 rows
Colour 5 – 2 rows
Colour 2 – 8 rows
Colour 3 – 12 rows
Colour 4 – 8 rows (RC 128)
Colour 5 – 10 rows
Colour 2 – 10 rows
Colour 3 – 10 rows
Colour 4 – 2 rows
Colour 5 – 8 rows
Colour 2 – 12 rows
Colour 3 – 8 rows
Colour 4 – 10 rows
Colour 5 – 10 rows
Colour 2 – 10 rows
Colour 3 – 2 rows
Colour 4 – 8 rows
Colour 5 – 12 rows
Colour 2 – 8 rows
Colour 3 – 10 rows (RC 256).
COR, set carriage to plain knitting, col 1 (MY) only in feeder A, k 4 rows plain fabric.
WY in feeder A, k 20 rows. Knock off with open edge.

Fringes
Punch one line of a card, leaving 21 unpunched and punching the last three holes (see Punchcard Graph B, p. 77).
Cast on 99 sts, positioning them so that the card repeats 4 times and there is an odd three on the left of the bed – see Fig. 2, p. 77, finished fringe.
Start as for the main fabric as far as **, but keep the card locked on row one and k 32 rows in each colour (128 rows in all). K off in WY and knock off the machine.

Finishing
Machine-wash both fringe and fabric on hottest setting (cotton wash). It is the agitation that felts the wool as well as the heat.
Tumble dry if possible for maximum effect, adding tennis balls to agitate. The pieces will be tangled. Snip waste knots and smooth out fabric.
Press hard with a hot steam iron, or a damp pressing cloth, and a lot of pressure, ironing over the surface repeatedly to bond the thread well together until the fabric is firm to the touch. Block it to the correct size at the same time.
Lay the fabric wrong side up on a flat surface, and with the blunter point of a *sharp* pair of scissors, cut the floats (loops of yarn) carefully up the diagonals. Leave the narrower stripes with the plain cut floats and, on the wider stripes, trim a small edge off each side of the cut float, so that the background shows through. Trim waste fabric and yarn ends; fold in half to make a square; mark fold line and cut in two.
Cut fringe floats close to the right-hand 3 st Fair Isle panels, and trim off the right-hand background fabric close to the bases of the fringes, but leaving 5mm ($\frac{1}{4}$in) to the left of the Fair Isle panels to give a hem to sew with (see Figs 1 and 2 on p. 77). The odd end will need to be opened out and the background hem allowance left on the uncurled background fabric.
Sew up with a short, narrow zigzag stitch on a sewing machine, inserting the fringes between

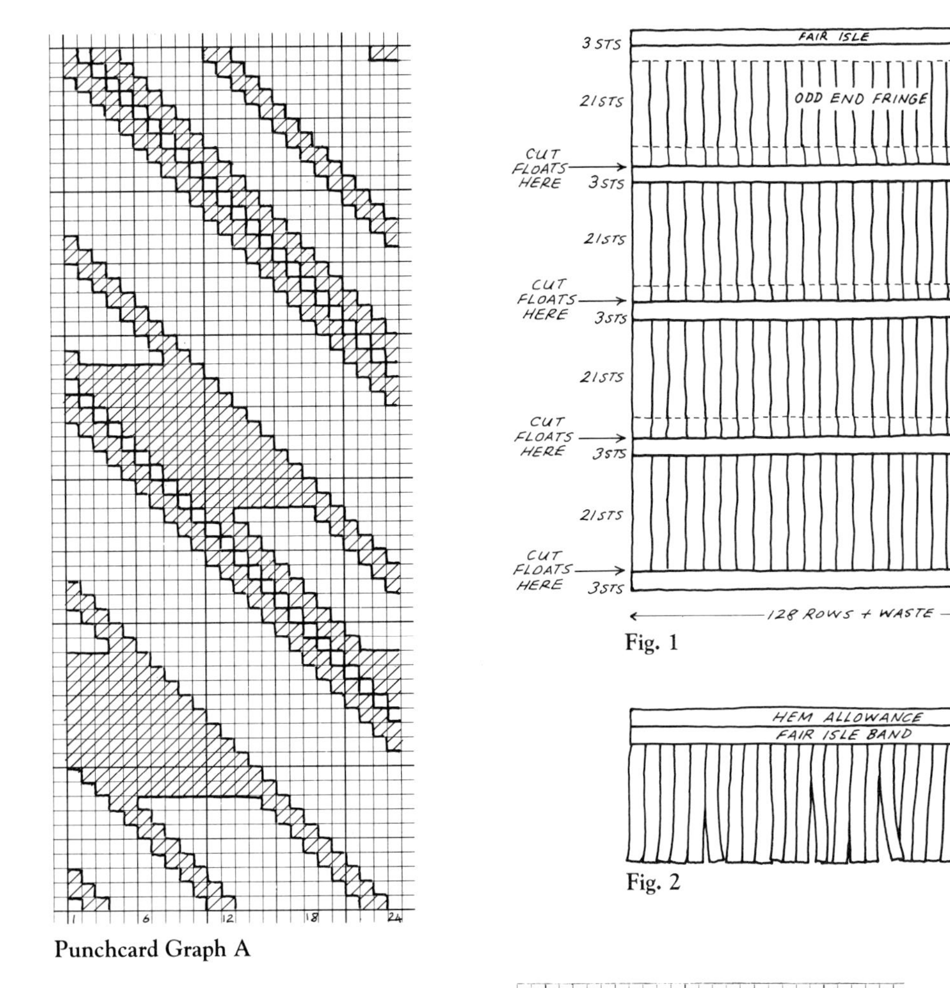

Punchcard Graph A

Fig. 1

Fig. 2

Punchcard Graph B

the seams as you go. Start with the zip insertion, laying the fringe between the right side of the fabric at the end that has the four rows of plain knitting, and the right side of one of the zip tapes. Pin, tack and sew. Next, place the other zip tape right side to right side with the end of the other piece that has four rows of plain knitting. Pin, tack and sew.

Do the 'fold' seam next and finally the two side seams, inserting the fringe as you go, and folding the ends of the fringes in to neaten.

Turn the cover through the zip – insert the cushion. Lightly steam and pull out the corners and fringe.

Fritillary Jumper

BY SOPHIE SCHELLENBERG

This simple machine knitted jumper achieves its three-dimensional effect through flowers and dragonflies which are knitted separately and then attached to the fabric. This could be tackled by an imaginative beginner.

MATERIALS

1 × 500g cone and oddments.
2/8 Shetland is the ideal yarn because of the range of subtle colours which blend so well. However, with changes in tensions any yarn may be used.

MEASUREMENTS

Actual chest size 106cm (42in)
Length at centre back 68.5cm (27in)
Sleeve seam 53cm (21in)

TENSION

8 stitches = 2.5cm (1in)
11 rows = 2.5cm (1in)
Shetland can be knitted on a tight tension to give it body.
Main tension T6
Fair Isle T7
Mock rib T1.1
Neckbands T4
3D additives T4

ABBREVIATIONS

carr–carriage, **CC**–Contrast Colour, **dec**–decrease, **inc**–increase, **k**–knit, **MC**–Main Colour, **N**–needle, **rep**–repeat, **st(s)**–stitch(es), **WY**–waste yarn.

Front

Cast on 85–0–84 after rib 85–0–85.
Mock rib 1 × 1 rib 2 Ns out on either end for ease of sewing up.
Cast on WY.
MC: 1 row T10, 29 rows T1.1, 1 row T10, 3 rows T1.1.
CC 1: 2 rows T1.1.
CC 2: 1 row T1.1.
MC: 24 rows T1.1.
Pick up onto empty Ns.
A ribber may be used but with Shetland yarn a mock rib gives more body.
4 rows MC: T6.
16 rows Fair Isle fritillary flower: T7. See chart, p. 80.
10 rows Fair Isle fritillary leaf. See chart, p. 80.
4 rows MC.
Rep Fair Isle.
K MC to row 230.

Front Neck

85–0–13 on hold carr on right.
K 1 row.
Dec 1 st on neck edge every row 5 times.
Dec 1 st every other row – 25–85.
K to row 266, last row T10, cast off.
Work left side as right.
Pick up neck band – the neck can either be knitted in plain knitting or rib.
MC: 7 rows T4.
CC 2: 1 row.
CC 1: 2 rows.
MC: 3 rows.
MC: 1 row T10.
MC: 13 rows T4.
Pick up and cast off.

Back

As front to row 250.

Back Neck

88–0–18 on hold.
K 1 row.
Dec 1 st every row 5 times.

Dec 1 st every other row – 25–85.
K to row 266.
Work left side as right.
Neck band as front.

Sleeves
39–0–38 after rib 39–0–39.
K as for body cuff – Fair Isle, and so on.
Inc 1 st each side every 3rd row to the end of the Fair Isle.
Inc 1 st each side every 4th row.
Total rows 180 approximately 86–0–86 stitches.

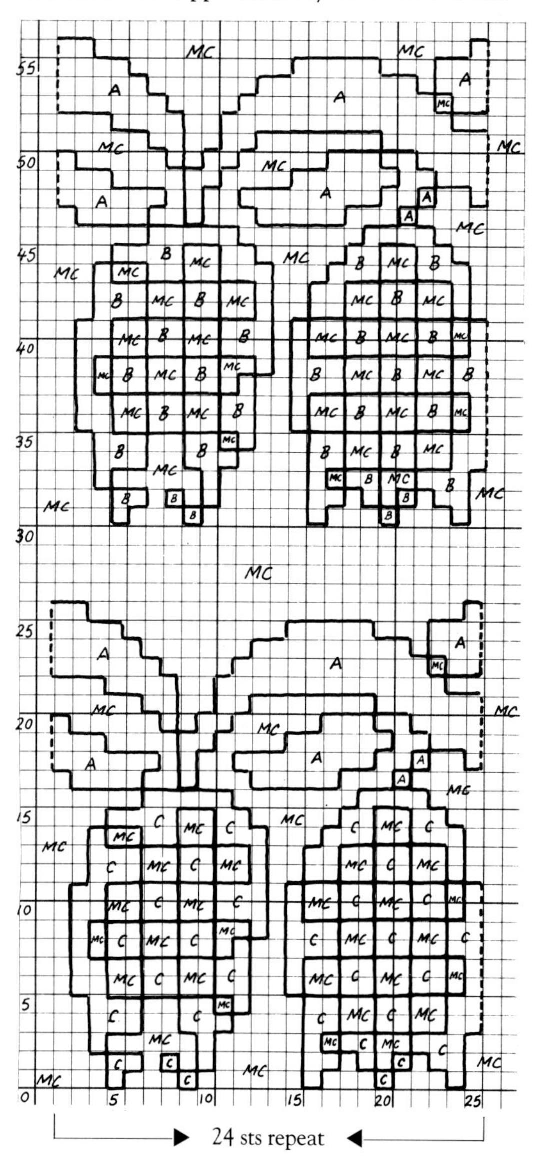

3D Additives
Cast on using 1 × 1 row then put all the Ns in working position so the flowers can be gathered easily. Use tension 4 for all 3D additives.
Two large leaves 5 sts 11 rows, 6 sts 9 rows, 7 sts 11 rows. Dec 1 st every 6th row until 3 sts remain. Dec 1 st every 3rd row.
Two small leaves 3 sts 5 rows, 4 sts 7 rows, 3 sts 5 rows, 2 sts 3 rows, 1 st 2 rows.
Two stalks 1 stalk 3 sts 140 rows, 1 stalk 3 sts 110 rows.
Two fritillary flowers Cast on 28 stitches.
Fair Isle – checks are best done manually.
2 sts in, 2 sts out, 2 sts in, 2 sts out, rep throughout flower pattern.
12 rows.
Cast off 5 sts each side, 12 sts on hold. K 2 rows.
Dec 1 st each side. K 2 rows.
Rep until 1 st remains, pull yarn through.
Rep with the next 6 sts and the last 6 sts.
Dragonfly 2 wings 3 sts 26 rows; 2 wings 3 sts 32 rows; 2 bodies 7 sts 4 rows 1 colour for head.
6 rows in a separate colour – continue with this for the tail. Dec 1 st every 7th row alternate sides to 3 sts, 3 rows dec 1 st and so on.
Use artistic licence when stitching on your 3D additions – it does take time and patience.

Choose three contrasting colours to make up the fritillary pattern. Use the same colour for each repeat of the leaf pattern but use a contrasting colour for the repeat of the flowers.

A = colour 1
B = colour 2
C = colour 3
MC = Main Colour

CROCHET

In its heyday in the Victorian period, most women were accomplished at crochet, and used it to make many household items. Its popularity declined during the early part of this century, it enjoyed a revival in the sixties and now, in the nineties, has made a comeback. Not only is crochet used to make stunning garments but it is also used increasingly as part of hand and machine knitted clothing.

There are many avid crochet fans eager to make the most of any new designs. In response the book includes some of the most well-loved and respected crochet designers in Britain today.

Lesley Conroy has been an avid enthusiast of crochet since she was five and this is revealed in her distinctive designs. Julia Jones brings back memories of the sixties with her bright fun top and Linda Parkhouse shows the versatility of crochet with her warm chunky jacket. Mary Konior is a leading specialist in the traditional art of filet crochet and her work is very intricate and beautiful.

Yoked Mungo Jacket

BY LESLEY CONROY

This unusual jacket is worked in fine yarns to give a light, soft feel. It is a fairly easy piece to make and suitable for a beginner to crochet. You could also make a hat to go with the jacket with the left over yarn.

MEASUREMENTS

Chest 122cm (48in) actual measurement
Length 79cm (31in)
Sleeve 43cm (17in) or desired length

MATERIALS

Philosopher's 2-ply pure wool as follows:
Main colour – 5 × 50g skeins
Contrast colours A, B, C, D and E – 2 × 50g skeins
(Leftover yarn is sufficient to also make a baby jacket or an adult accessory such as a hat.)
8 buttons
Crochet hooks, sizes 4mm and 3.5mm.

TENSION

14.5 tr to 10cm (4in) over body pattern on 4mm hook.
16 tr to 10cm (4in) over yoke pattern.
8 rows to 10cm (4in) over all patterns.
If there are too few sts to 10cm (4in), use a finer hook; if there are too many sts to 10cm (4in), use a coarser hook.
N.B. Turning chain counts as a stitch throughout.

ABBREVIATIONS

alt–alternate, **ch**–chain, **ch sp**–chain space, **cont**–continue, **dc**–double crochet, **dc2tog**–double crochet 2 together, **dec**–decrease, **foll**–follow(ing), **htr**–half treble, **inc**–increase, **MC**–main colour, **patt**–pattern, **rem**–remain(ing), **rep**–repeat, **rtrb**–raised treble back, **rtrf**–raised treble front, **ss**–slip stitch, **st(s)**–stitch(es), **st ch**–starting chain, **tch**–turning chain, **tog**–together, **tr**–treble, **tr tr**–triple treble, **tr2tog**–treble 2 sts together, **yrh**–yarn round hook

NOTES

1 To change colour, work to stage where there are two loops left on the hook, drop the old colour and pick up new, yrh and draw through last 2 loops on hook. This changes colour neatly, with no colour drag.
2 Work stripes of 2 rows each, working colours in a random way as you like.
3 To work the triple treble into corresponding st 2 rows below, make an uncompleted tr into next st (i.e. to stage where there are two loops left on hook), then yrh 3 times, then insert the hook around the stem of the st two rows below from front to back and round to the front again, yrh and draw through a loop (yrh and draw through 2 loops) 3 times, then yrh and draw through 3 loops.
4 To work corded edging, work 1 row of double crochet from left to right through front loop only.

Pocket Linings

With 4mm hook and MC, make 18 ch.
Row 1 Work 1 tr into 4th ch from hook and into each ch to end, turn.
Row 2 3 ch (counts as first tr), 1 tr into each tr to end, turn.
Rep row 2 until pocket lining measures 10cm (4in). Fasten off.

Right Half of Body

Using a 4mm hook and MC, make 72 ch. Work in patt as follows:
Row 1 (Right side) 1 tr in 4th ch from hook, 1 tr in each ch to end. 70 tr. Turn.
Row 2 3 ch (counts as 1 st tr), 1 tr into 2nd tr and every tr in row, change colour, turn.

Row 3 3 ch, * 1 tr tr into corresponding st 2 rows below, 1 tr into next st, rep from * to last 3 sts, 1 tr tr, 2 tr, turn.
Row 4 As row 2, changing colour at end of row, turn.
Row 5 3 ch, 1 tr, * 1 tr tr into corresponding st 2 rows below, 3 tr, rep from * to end, turn.
Row 6 As row 2, changing colour at end of row, turn.
Row 7 3 ch, *(1 tr tr into corresponding st 2 rows below) 3 times, 5 tr, rep from * to last 5 sts, (1 tr tr into corresponding st 2 rows below) 3 times, 2 tr, turn.
Row 8 As row 2, changing colour at end of row.
Rows 3 to 8 form patt. Rep these rows changing colours as desired every 2 rows.
When work measures 25cm (10in), shape for armhole as follows:
If on an even row leave last 4 tr unworked. If on an odd row, ss across first 4 tr and into 5th tr, 3 ch, patt to end.
Then work straight until work measures 30cm (12in), ending with an even-numbered row, then place pocket as follows:
Patt 31 sts, work across 16 tr of pocket lining, leaving 16 tr of body, patt across rem body sts.
Cont in patt until work measures 36cm (14in), ending with an even numbered row, then shape for armhole by making 6 ch, turn.
Next row Make 1 tr into 4th ch from hook and into next 2 ch, then patt to end. When work measures 60cm ($23\frac{1}{2}$in), i.e. after 47 rows, fasten off. You must finish on an odd-numbered row.

Left Half of Body
Work left half of body as for right half, reversing all shaping, so that left half matches right. Do not fasten off. Join to right half of body with a ss seam on wrong side of the work. Fasten off.
(N.B. Make pocket opening 19 sts from bottom edge of body. For armhole leave last 4 sts of row 21 unworked. When increasing for armhole, with colour to be used on row 29, ss into last st, make 4 ch, fasten off. When working row 29, work into the 4 ch just made.)

Back Yoke
With MC and with right side facing, rejoin yarn to top of back at armhole edge, with a ss. 1 ch (counts as first dc), work 76 dc evenly across top of back. Do not turn, work 1 row of corded edging, working through the front loop only of dc. Change colour.
Row 1 3 ch, work 1 tr in back loop of each dc to end. Turn (77 tr).
Row 2 3 ch, work 1 tr in each st to end, change colour, turn.
Row 3 3 ch, * 1 tr tr in corresponding st of row below, 1 tr in next st, rep from * to end, turn.
Row 4 As row 2, changing colour at end of row.
Row 5 3 ch, * 1 tr, 1 tr tr in corresponding st 2 rows below, rep from * to last 2 sts, 2 tr, turn.
Row 6 As row 2, changing colour at end of row.
Rows 3 to 6 form the patt. Rep these rows until yoke measures 25cm (10in), ending with a right side row. Fasten off.

Right Front Yoke
With right side facing and MC, work 39 dc evenly across right front. Work 1 row corded edging as given for back, through front loops only. Work in yoke patt as given for back yoke, dec 1 st at neck edge by tr2tog on 2nd row and then every row at neck edge until 23 tr remain. Then work straight in patt until yoke front matches yoke back. Work

left yoke front in same way, reversing shapings. Join shoulder seams by working 1 row dc with MC through back and front, working from the right side of the work. Do not turn, then work 1 row corded edging. Fasten off.

Sleeves
With right side facing and MC, rejoin yarn and work 77 dc evenly around armhole. Work 1 row corded edging as given for back through front loops of dc only. Change colour.
Work in patt as given for back.
When work measures 5cm (2in), begin to work in rounds by joining last st to first with a ss. Cont to work in patt in rounds, turning after every round. When sleeve measures 9cm (3.5in) dec 1 st at each end of next and every alt round 10 times (57 tr). Then dec 1 st at each end of every round 6 times (45 tr).
Work straight until sleeve measures 35cm (14in), or desired length, from the start of working in rounds. Work 1 round dc, dec to 32 sts. Work 8cm (3in) in treble rib with a 3.5mm hook, working random stripes if desired. Work treble rib as follows:
Round 1 (Wrong side) 2 ch (stand as rtrb), * 1 rtrb around stem of next st, 1 rtrf around stem of next st, rep from * to end, ss into top tch, turn.
Round 2 (Right side) 2 ch (stand as rtrb), * 1 rtrf around stem of next st, 1 rtrb around stem of next st, rep from * ending 1 rtrf around stem of last st, ss into top tch, turn. Fasten off.

Welt
Using 3.5mm hook, with right side facing, work 175 dc along bottom edge of jacket, working approx 2 dc into every row end 5 times, and 1 dc into every 6th row end. Work in raised treble rib as follows:
Row 1 (Wrong side) 2 ch, * 1 rtrb around stem next st, 1 rtrf around next st, rep from * ending 1 rtrb around stem last st, 1 htr into top tch, turn.
Row 2 2 ch, * 1 rtrf around next st, 1 rtrb around next st, rep from * ending 1 rtrf around last st, 1 htr into top tch.
Rep rows 1 and 2 until rib measures 8cm (3in), working random stripes if desired, or working all rib in MC if desired. Change to a 4mm hook if you feel the rib is pulling in too much. Do not fasten off.

Front Bands and Shawl Collar
Using 3.5mm hook and MC, work 3 dc for every 2 row ends of rib, work 1 dc in each st of body to yoke, place marker, work 2 dc to every row end to shoulder, work 1 dc in every st across the back, working dc2tog at back neck corners, work 2 dc to every row end to end of yoke, place marker, work 1 dc in each st along front to top of rib, work 3 dc for every 2 row ends of rib, turn.
Next row 1 ch, work in dc to marker, work in treble to next marker, work in dc to end, turn.
Next row 1 ch, work in dc to marker, work in tr rib to shoulder, turn. (N.B. After last tr rib st, work a dc, then turn).
Next row Work in tr rib to shoulder, turn.
Next row Work in tr rib for 5 extra sts, turn.
Next row Work in tr rib for 5 extra sts, turn.
Cont in this way, turning and working 5 extra sts in tr rib on every row until markers are reached, then cont in dc to end of row. Work 3 extra rows, working eight buttonholes on right front (for a woman), left front (for a man), on next row. Make buttonholes by working to buttonhole position, 1 ch, miss 2 dc, work in dc to next buttonhole position. On next row, work 2 dc in 1 ch sp of buttonhole. Remember to continue working the shawl collar in tr rib on these rows. You might like to work the last row in a contrast colour.

Pocket Edgings
With 3.5mm hook and right side facing, rejoin yarn to pocket top. 1 ch, work 1 dc into each of next 15 st (16 dc). 1 ch, do not turn. Work 1 row corded edging through both loops. Fasten off. Catch st neatly to jacket.

Finishing
Weave in all ends. Sew on buttons to correspond with buttonholes.

Fuchsia Pink Camisole

BY JULIA JONES

This bright fun camisole in pure cotton is very quick to work and could be made by anyone with basic crochet skills. It makes a lovely piece of simple evening-wear which can be dressed up or down according to the occasion.

MEASUREMENTS
To fit 91–97cm (36–38in) bust
Length at centre back approx 60cm (23in)

MATERIALS
Rowan Cabled Mercerized Cotton in Fuchsia (Shade 326) 9 × 50g balls
4mm crochet hook; tapestry needle

TENSION
6 sts and 6 rows to 2.5cm (1in) worked over main dc pattern

ABBREVIATIONS
beg–beginning, **ch**–chain, **ch sp**–chain space, **cont**–continue, **dc**–double crochet, **dec**–decrease, **foll**–follow(ing), **gr**–group, **inc**–increase, **patt**–pattern, **rep**–repeat, **ss**–slip stitch, **st(s)**–stitch(es), **tr**–treble

Lower Border
Beg with 20 ch, turn.
Row 1 Miss 5 ch, into 6th ch from hook work 1 tr* 1 ch, miss 1, 1 tr, rep from * 6 times, turn.
Row 2 1 dc in every st of row 1, inserting hook *only into back thread of previous dc*, 3 ch, turn.
Row 3 Miss 1 ch, 1 dc into next 2 ch, 1 dc into 1st st of dc, 5 tr into next st (this forms the 'tuft'), 15 dc, 1 ch turn.
Row 4 15 dc over dc of previous row, 1 tr behind the 'tuft' into a st on previous row, 3 dc, 3 ch, turn.
Row 5 Miss 1 ch, 1 dc into each of next 2 ch, 1 dc into 1st dc of previous row, a 'tuft' of 5 tr into next st, 3 dc, a 'tuft' into next st, 13 dc, 1 ch, turn.

Row 6 13 dc, 1 tr, 3 dc, 1 tr, 3 dc, 3 ch.
Row 7 Miss 1 ch, 1 dc into each ch, 1 dc into 1st dc of previous row, a 'tuft', 3 dc, a 'tuft', 3 dc, a 'tuft', 11 dc, 1 ch, turn.
Row 8 11 dc, 1 tr, 3 dc, 1 tr, 3 dc, 1 tr, 1 dc, leave 2 sts unworked, 1 ch turn.
Row 9 3 dc, a 'tuft', 3 dc, a 'tuft', 13 dc, 1 ch turn.
Row 10 13 dc, 1 tr, 3 dc, 1 tr, 1 dc, leave 2 sts unworked, 1 ch turn.
Row 11 3 dc, a 'tuft', 15 dc, 1 ch, turn.
Row 12 15 dc, 1 tr, 1 dc, leave 2 sts unworked, 1 ch, turn.
Row 13 4 ch, 1 tr into 3rd st of dc, * 1 ch, miss 1, 1 tr, repeat from * 6 times.
Rep from row 2 until work measures 102cm (40in) (16 patt repeats).

Main Body of Garment (One Piece)
Break off yarn and rejoin at top edge of border patt. Work 1 row of 268 dc, ss final dc into 1st dc of this row to form a circle. Thread a tapestry needle with yarn and ss border edges to join.
Now work in rounds of dc, *always working only into back thread of each st*, until the work measures 40cm (15½in) approximately.
The work now splits to form sleeve openings. Mark side seam with a spare length of coloured yarn and turn work.
Shape front Rejoin yarn 10 dc from marker thread (including marked st) and proceed as folls:
1st row 1 ch, 114 dc (*worked in back thread of each st*), 1 ch, turn.
2nd row Work 1 dc as before into each dc to last 4 dc, 1 ch, turn.
3rd row As 2nd row.
Continue to dec 4 sts on each row in this way until 6 decs in all have been made.
Work rows of dc on these sts without further decs until work measures 12cm (4½in) from start of armhole, finishing at right outer edge of work.
Next row 30 dc, 1 ch, turn.
Continue in patt, dec 1 st at neck edge on every 2nd row until 16 st remain. Fasten off. Rejoin yarn and complete left side, reversing shapings.
Shape back To work back miss 10 dcs from marker yarn, rejoin yarn and work 114 dcs. Continue as for front, matching shapings. (A space of 20 dcs is left between front and back armhole shapings.)

To Make Up
Join shoulder seams and press lightly on the wrong side with a steam iron, pulling points into place on lower border.

Neck and Armhole Borders
Join yarn to lower edge of armhole border and work 146 dc around armhole. Work 1 more row of dc as before, working into back thread only of each previous st. Ss into 1 st dc of last row. Fasten off. Repeat for other armhole.
Work a border in the same manner around neck edge, working 36 dc along front edge, 52 dc over shoulder, 36 dc along back edge and 52 dc down second shoulder. Work another row of dc, ss into 1st st and fasten off. Press lightly with a steam iron.

Collar in Filet Crochet

BY MARY KONIOR

This charming cotton collar is worked in filet crochet using a fine hook and very fine thread. The work must always be kept very clean and the tension even. It gives a beautiful finishing to many styles of dress and blouse.

Filet crochet designs were based on filet lace, also known as lacis, and of ancient origin. Filet crochet usually looks best worked in a fine crochet cotton such as No. 60. It should be worked at a firm, even tension. Use the smallest hook you can manage, but if your hook continually splits the thread, it is too small. When well-made in good quality cotton, filet crochet holds its shape well and can be machine washed. It should not need starching. This collar is worked in short rows widthways so that its length can be adjusted as required. The curvature will ease to fit most necklines and, if preferred, ribbon can be threaded at the neck edge. Alternate rows increase at the neck edge and this unusual method of shaping gives a diagonal look to the rows.

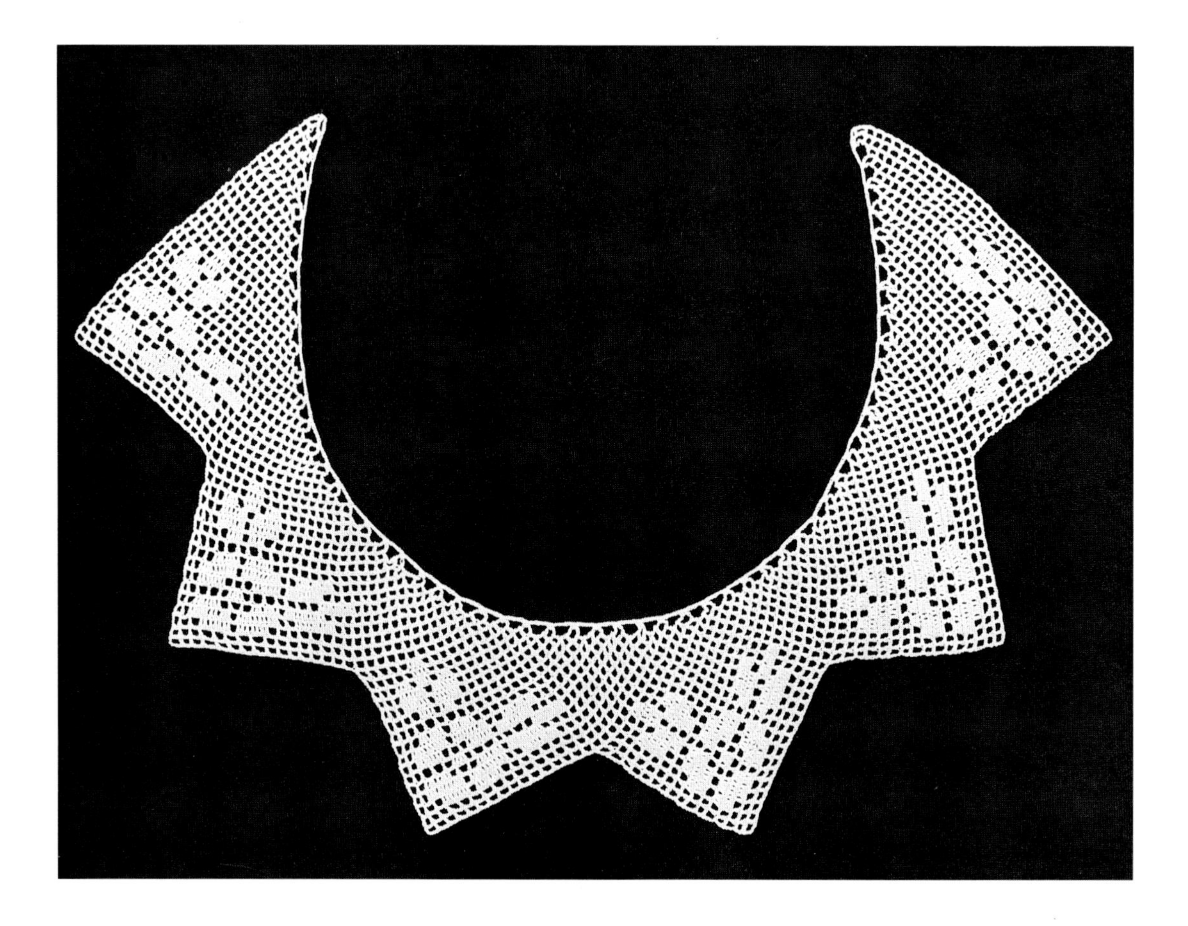

MATERIALS

One 20g ball of Coats Mercer Crochet No. 60
Steel crochet hook size 0.75mm

MEASUREMENTS

Depth 8cm (3in)
Length 51cm (20in) or as required, curvature adjustable

To Work the Collar

Start at the base of Fig. 1, with 11 chain as shown. When the top of the diagram is reached, continue from the 22nd row of Fig. 2.
Repeat from A to B four more times, or as required. Then work along the neck edge with, * 3 chain, 1 double crochet into next chain loop. Repeat from * all along. Finish off.
Press with a steam iron, and thread ribbon through the neck edging if required.

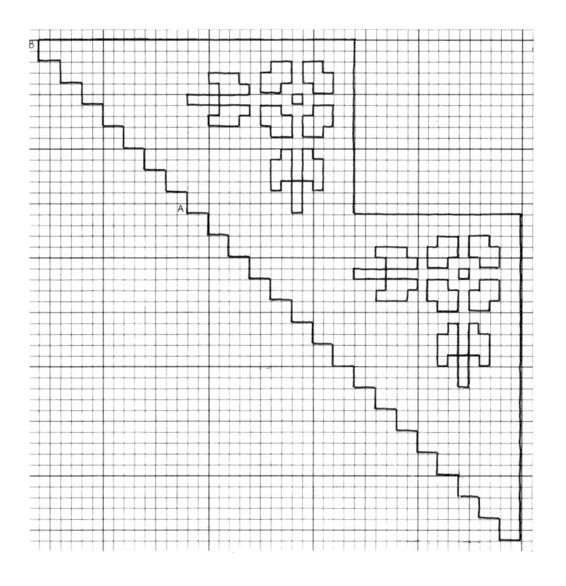

Fig. 2

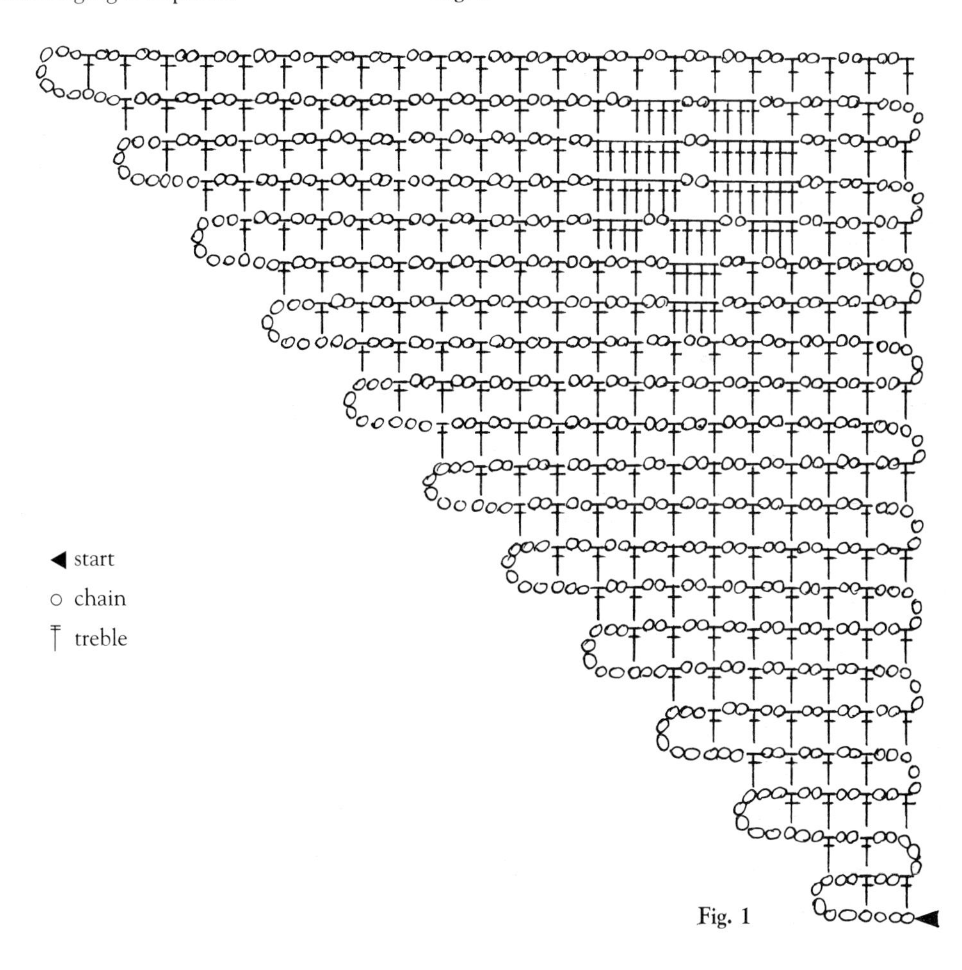

Fig. 1

Ladies' Crochet Jacket

BY LINDA PARKHOUSE

This winter-weight jacket is extremely quick to make – it could even be finished in just a couple of evenings. But note that care should be taken with the making-up of the garment.

MEASUREMENTS

To fit chest 81(87,91,96)cm/32(34,36,38)in
Actual width round chest at underarms
102(104,109,111)cm/40(41,43, $43\frac{3}{4}$)in
Length from shoulder to hem
74(76,78,80)cm/29(30,$30\frac{1}{2}$,$31\frac{1}{2}$)in
Sleeve seam 45(45,46,46)cm/$17\frac{3}{4}$($17\frac{3}{4}$,18,18)in

MATERIALS

Sirdar's Wash'n'Wear Super Chunky in Cherry – 15(16,17,18) × 100g balls
Crochet hooks size 7mm, 6mm and 5.5mm
14 buttons of 2cm ($\frac{3}{4}$in) diameter

TENSION

5 complete patts and 9 rows to 10cm (4in) over patt, using 7mm hook.

ABBREVIATIONS

alt–alternate, **beg**–begin(ning), **ch**–chain, **cont**–continue, **dc**–double crochet, **dec**–decreasing, **foll**–follow(ing), **inc**–increasing, **patt**–pattern, **rem**–remaining, **rep**–repeat, **ss**–slip stitch, **sp**–space, **st(s)**–stitches

Back

With a 7mm hook, work a foundation row of 52(54,56,58) ch.
1st row (1 dc, 1 ch, 1 dc) into 3rd ch from hook, * miss 1 ch, (1 dc, 1 ch, 1 dc) into next ch, rep from * to last ch, 1 dc into last ch, 2 ch to turn. 50(52,54,56) sts and 25(26,27,28) patts.
2nd row * (1 dc, 1 ch, 1 dc) into ch sp, rep from * to end, 1 dc into turning ch, 2 ch to turn.
Rep row 2 only to form patt.
Work straight until piece measures 38(40,42,44)cm/15($15\frac{3}{4}$,$16\frac{1}{2}$,$17\frac{1}{4}$)in from beg, ending a wrong-side row.

Back Yoke

Work as for back for 32cm ($12\frac{1}{2}$in), ending a wrong-side row. Mark centre 7(8,9,10) patts for back neck with contrasting thread.

Fronts (work 2 pieces alike)

With 7mm hook, work a foundation row of 26(26,28,28) ch.
Work in patt as for back on these 12(12,13,13) complete patts until piece measures 38(40,42,44)cm/15($15\frac{3}{4}$,$16\frac{1}{2}$,$17\frac{1}{4}$)in from beg, ending a wrong-side row.

Right Front Yoke

With 7mm hook, work a foundation row of 26(26,28,28) ch, then work as for front sections until piece measures 25(24,23,22)cm/ $9\frac{3}{4}$($9\frac{1}{2}$,9,$8\frac{1}{2}$)in from beg, ending a wrong-side row.

Shape Front Neck

Ss across 4 sts at beg of next row, and into 5th st, then work in patt to end. Dec 1 st at neck edge on next and every foll row until 18 sts rem or 9 complete patts.
Work straight in patt on these sts until front yoke measures the same as back yoke, ending with a wrong-side row.

Left Front Yoke

Work as for right front yoke, reversing all shapings.

Sleeves
With 7mm hook, work a foundation row of 30 ch. Work in patt as for body piece to give 28 sts and 14 complete patts.
Shape side edges Inc 1 st at each end of next and every foll alt row 18 times in all, until 64 sts are gained and 32 complete patts.
Work straight in patt as set on these sts until sleeve measures 41(41,42,42)cm/16(16,16½,16½)in from beg, ending with a wrong-side row.

Cuffs (work 2 pieces alike)
With 6mm hook, work a foundation row of 7 ch.
1st row 1 dc into 2nd ch from hook, then work 1 dc into each ch to end. 6 dc, turn with 1 ch.
2nd row Miss 1 dc, * work 1 dc into back loop only of next dc, rep from * to end, turn with 1 ch.
Rep this 2nd row only to form patt.
Work straight until piece measures 26cm (10½in) from beg, ending with a wrong-side row.

Pocket Flaps (work 2 pieces alike)
Work as for cuffs for 12cm (4¾in) only.

Front Edgings
Buttonband Work as for cuffs for 69(70,71,72)cm/27(27½,28,28½)in, ending with a wrong-side row.
Buttonhole band Work as for buttonband for 4(5,2,3)cm/1½(2,¾,1¼)in, working and placing buttonholes as follows:
Buttonhole row 1 Miss 1 dc, work 1 dc into back loop only of next 2 dc, work 2 ch, miss 2 dc, work 1 dc into back loop only of each of next 2 dc, then turn with 1 ch.
Buttonhole row 2 Miss 1 dc, 1 dc into back loop only of next dc, work 1 dc into each of next 2 ch, then work 1 dc into back loop only of each of next 2 dc, turn with 1 ch. Cont in patt as set, working buttonholes at intervals of 7(7,7.5,7.5)cm/2¾(2¾,3,3)in until there are 10 worked, then complete as for buttonband.

Lower Edgings
Work as for cuffs for 96(98,103,105)cm/38(38½, 40½,41½)in, ending with a wrong-side row.

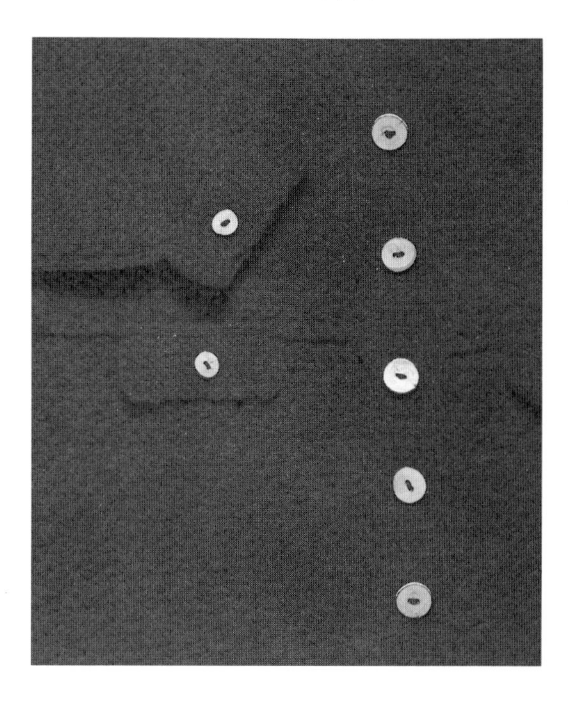

Collar
With 6mm hook, work a foundation row of 16 ch. Work in patt as for cuffs on these 15 sts for 55cm (21½in), ending with a wrong-side row.

To Make Up
With wrong sides together, and 5.5mm hook, work 1 dc row along shoulder edges of front and back yokes to make decorative seam on right side of work. Position front flaps centrally on fronts, and tack in place, then join fronts to yokes as for shoulder seams, and rep for back section. Fold sleeves in half lengthwise, and place fold to shoulder seam. Join sleeves to body, matching lower edge of sleeves to yoke seam. Join side and sleeve seams with flat seam. Join cuffs into ring and then, using flat seam, sew to cuff edge on sleeves, easing in fullness carefully. Join lower edging to body, matching front edges and easing fullness in along flat seam. Sew front edgings to respective front edges, using flat seam once again. Sew on collar to neck edges from middle of buttonhole band to middle of buttonband. Sew buttons to buttonband. Sew button centrally to each pocket flap through all thicknesses and trim cuffs with remaining buttons. Darn in loose ends. Do not press.

STITCHING

This section covers a number of needlework areas: needlepoint, cross-stitch, embroidery, machine embroidery, patchwork and rug-making. With the exception of machine embroidery, all of these needlecrafts have been practised for centuries. Here we have a fabulous collection of modern designs which really make the most of the stitching tradition from which they come.

Kaffe Fassett and Glorafilia present needlepoint projects which show off the stunning effects which can be achieved in colour and texture. The cross-stitch designs of Julie Hasler and Angela Wainwright will suit the skills of both the experienced and less experienced cross-stitcher. Lois Vickers and Linda McDevitt give very different examples of work which can be achieved with hand embroidery; while Paddy Killer, Gail Harker and Ariella Green show off the stunning and very individual embroidery achievements which are possible with a sewing machine. The patchwork designs of Freda Parker and Jane Walmsley elevate this age-old craft of recycling precious fabrics into an art form. Ann Davies gives a hooky rug design which can be simplified for beginners or given greater detail for experienced rug-makers.

Autumn Rose Cushion

BY KAFFE FASSETT

This is a beautiful cushion from one of the world's foremost needlepoint designers. But possibly of more interest is the fact that this work can be made by most people with basic needlepoint skills.

MEASUREMENTS

Finished size 44 × 45.5cm ($17\frac{1}{4}$ × 18in)

MATERIALS

Double thread canvas, 10 holes to 2.5cm (1in), 60cm (24in) square. Tapestry needle, size 18. Anchor tapestry yarn, as follows:

Shade	*Skein*	*Shade*	*Skein*	*Shade*	*Skein*
8006	3	9214	3	8546	1
8058	2	8884	3	8484	2
8014	1	8602	1	8938	2
8392	2	9452	10	9002	2
8442	2	8092	2	9206	3
8524	1	8022	1	8100	2
8522	1	8454	2	8064	7
8918	2	8438	3	8426	6
8962	3	8402	2		

Beginning the Work

Begin the work with a knot on the right side of the canvas and stretch the thread under the motif about to be worked for a distance of about 4cm ($1\frac{1}{2}$in). When this part of the work is completed and the stitches have anchored, the thread at the back can be cut off to neaten.

Half-Cross-Stitch

The half-cross-stitch is not as popular as the continental stitch but, in situations where a less thick fabric is needed, such as for pictures, it is very useful. It can be worked vertically or horizontally, but each stitch must be made in the same direction – top right to bottom left if you are working horizontally, or bottom left to top right if you are working vertically. When working in the hand it will be necessary to turn the work after each row. The numbers on the diagrams indicate the progression of the stitches as well as which way up the work is held to make them.

Continental Stitch (Tent stitch/Basketweave)

The long diagonal threads formed on the back of the canvas by this stitch create a padding of yarn that adds greatly to the durability of a piece. It can be worked vertically, horizontally and even diagonally, but like the half-cross-stitch, each stitch must be made in the same direction and it is necessary to turn the work after each row.

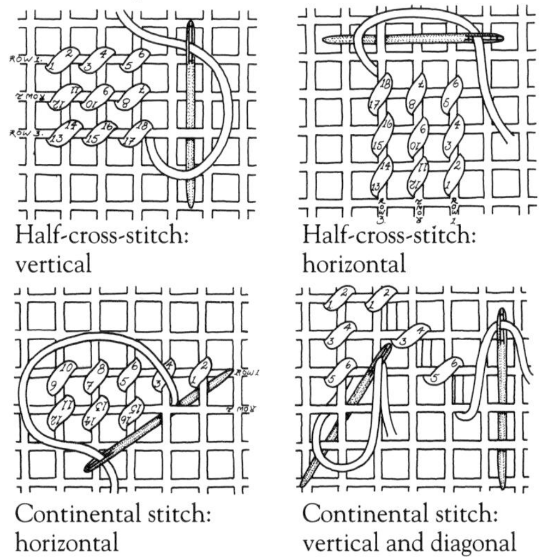

Half-cross-stitch: vertical

Half-cross-stitch: horizontal

Continental stitch: horizontal

Continental stitch: vertical and diagonal

Finishing off the Work

Finish off the work by bringing the end of the thread onto the right side before working the motif. Should there not be sufficient unworked canvas to use this method, turn the work the wrong side and run the thread through the worked canvas, once horizontally and once vertically, before cutting the thread.

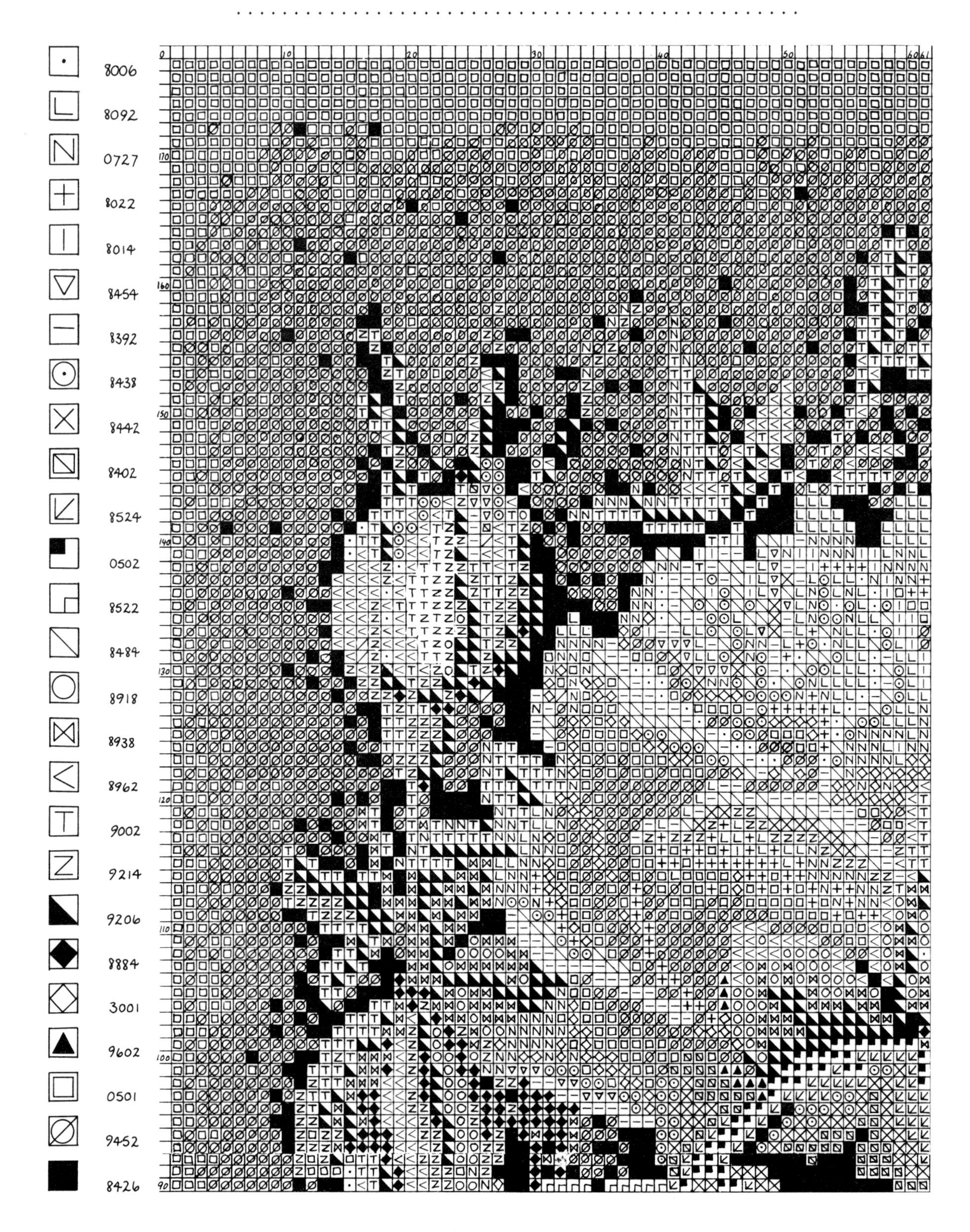
8006
8092
0727
8022
8014
8454
8392
8438
8442
8402
8524
0502
8522
8484
8918
8938
8962
9002
9214
9206
8884
3001
9602
0501
9452
8426
0
10
20
30
40
50
60
61
170
160
150
140
130
120
110
100
90

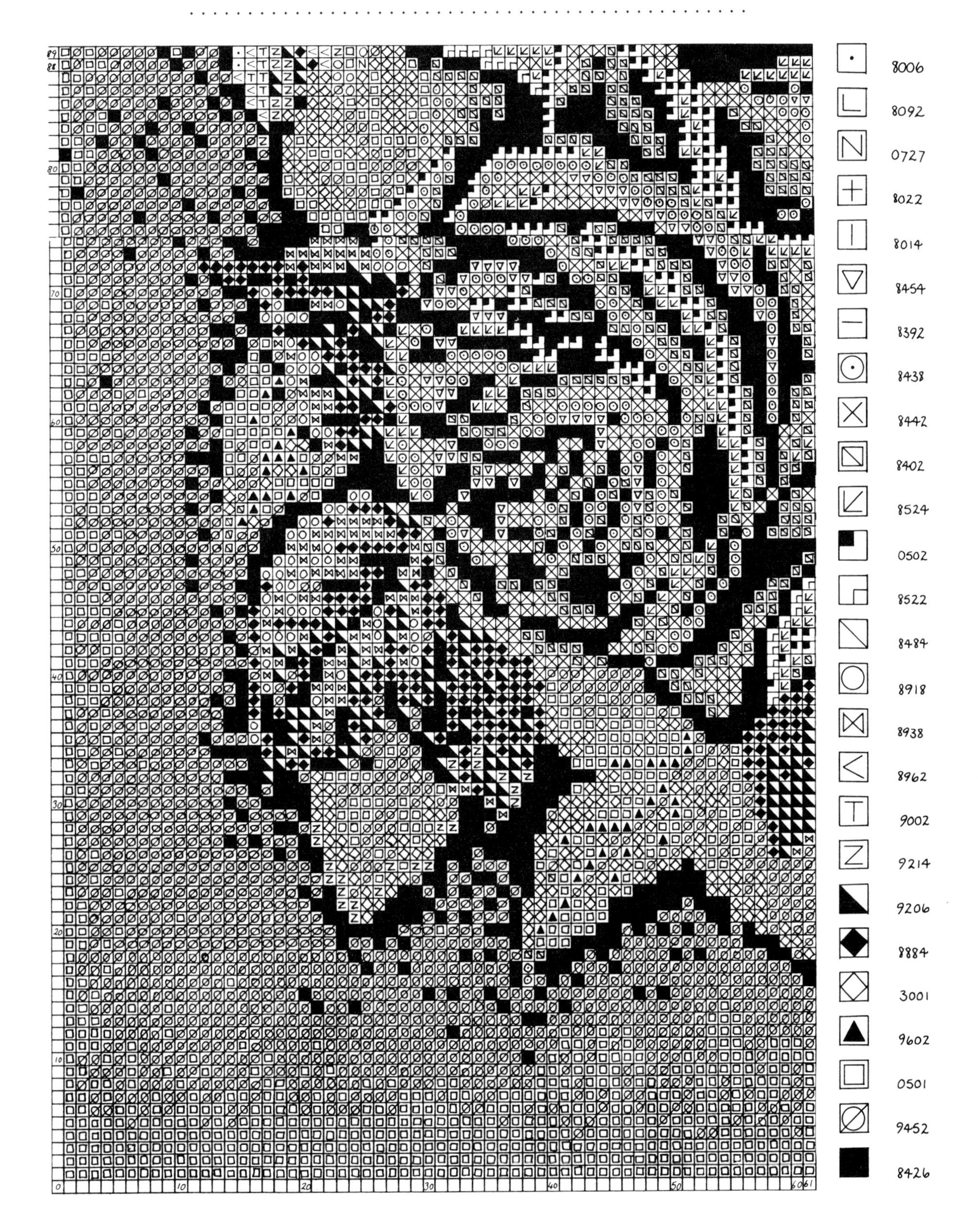
8006
8092
0727
8022
8014
8454
8392
8438
8442
8402
8524
0502
8522
8484
8918
8938
8962
9002
9214
9206
8884
3001
9602
0501
9452
8426

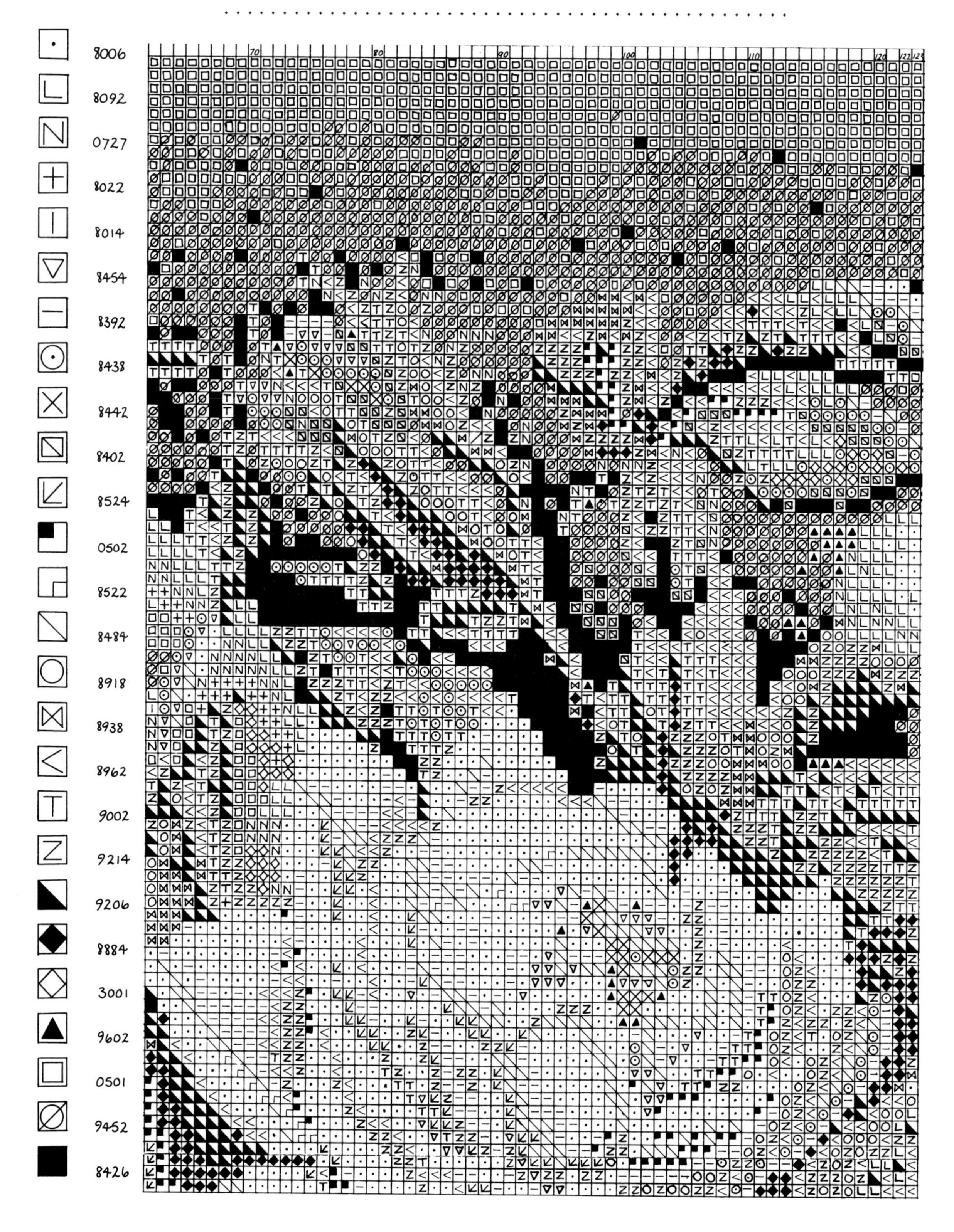
8006
8092
0727
8022
8014
8454
8392
8438
8442
8402
8524
0502
8522
8484
8918
8938
8962
9002
9214
9206
8884
3001
9602
0501
9452
8426
70
80
90
100
110
120
122
123

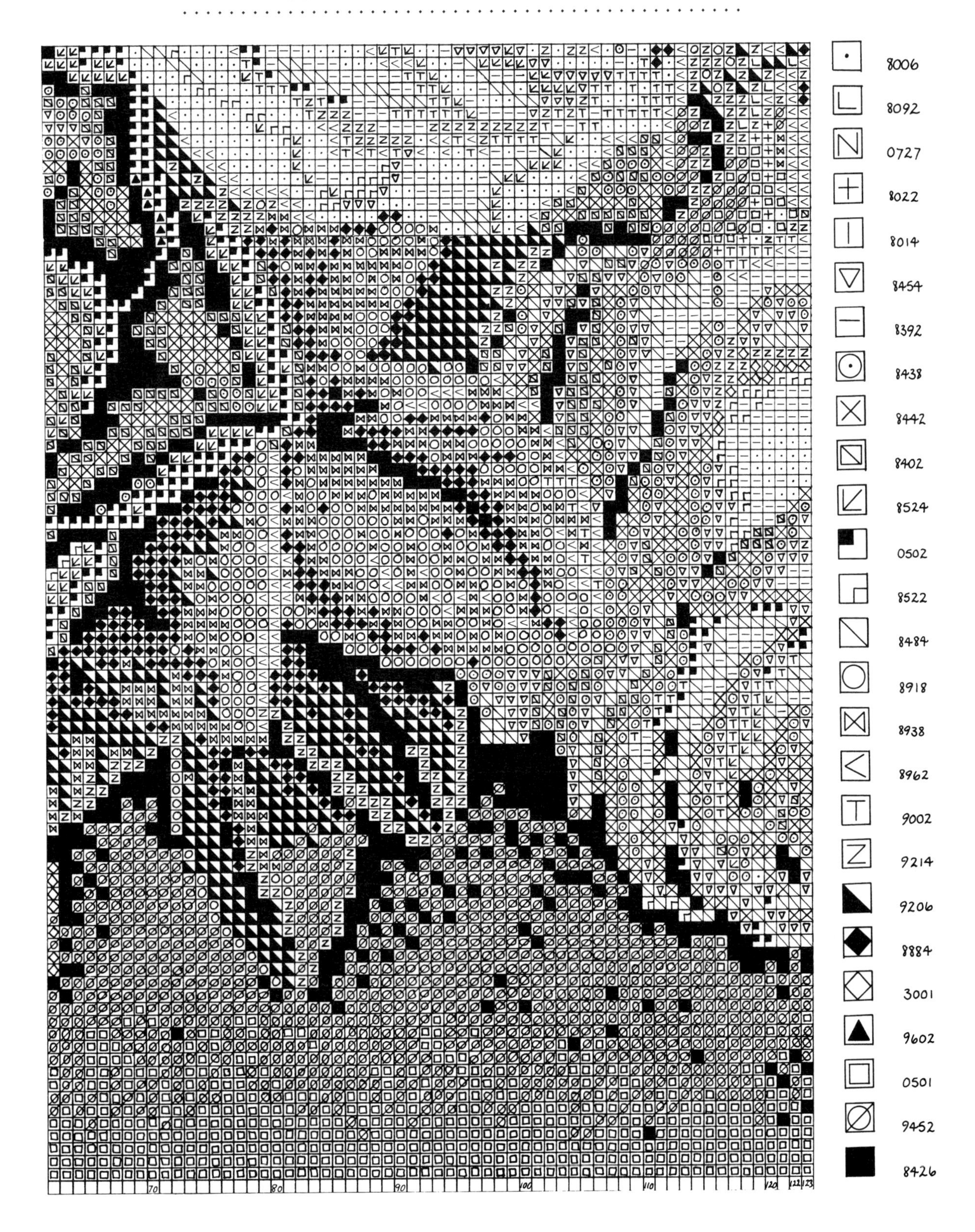
8006
8092
0727
8022
8014
8454
8392
8438
8442
8402
8524
0502
8522
8484
8918
8938
8962
9002
9214
9206
8884
3001
9602
0501
9452
8426
70
80
90
100
110
120
122
123

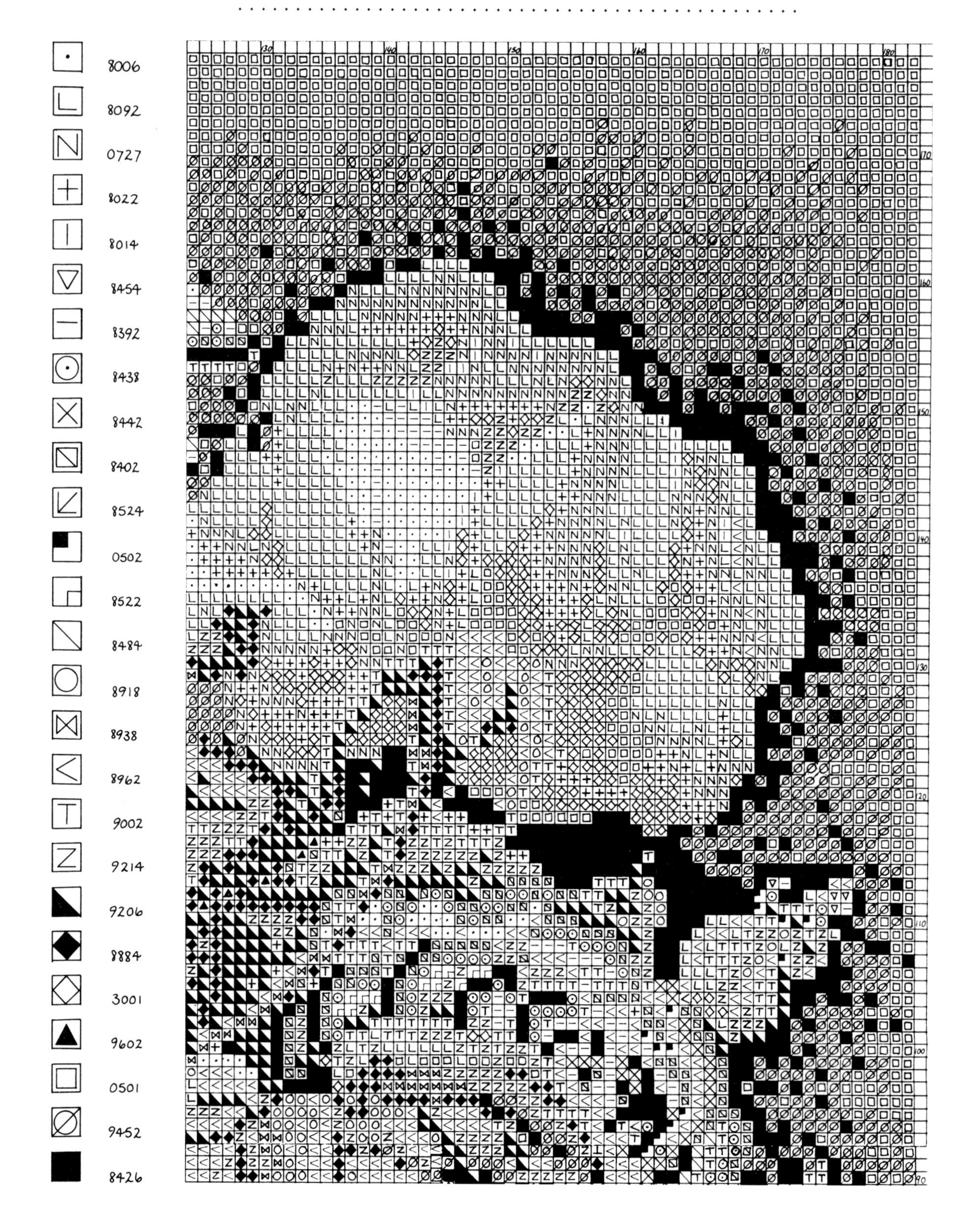
8006
8092
0727
8022
8014
8454
8392
8438
8442
8402
8524
0502
8522
8484
8918
8938
8962
9002
9214
9206
8884
3001
9602
0501
9452
8426
130
140
150
160
170
180
170
160
150
140
130
120
110
100
90

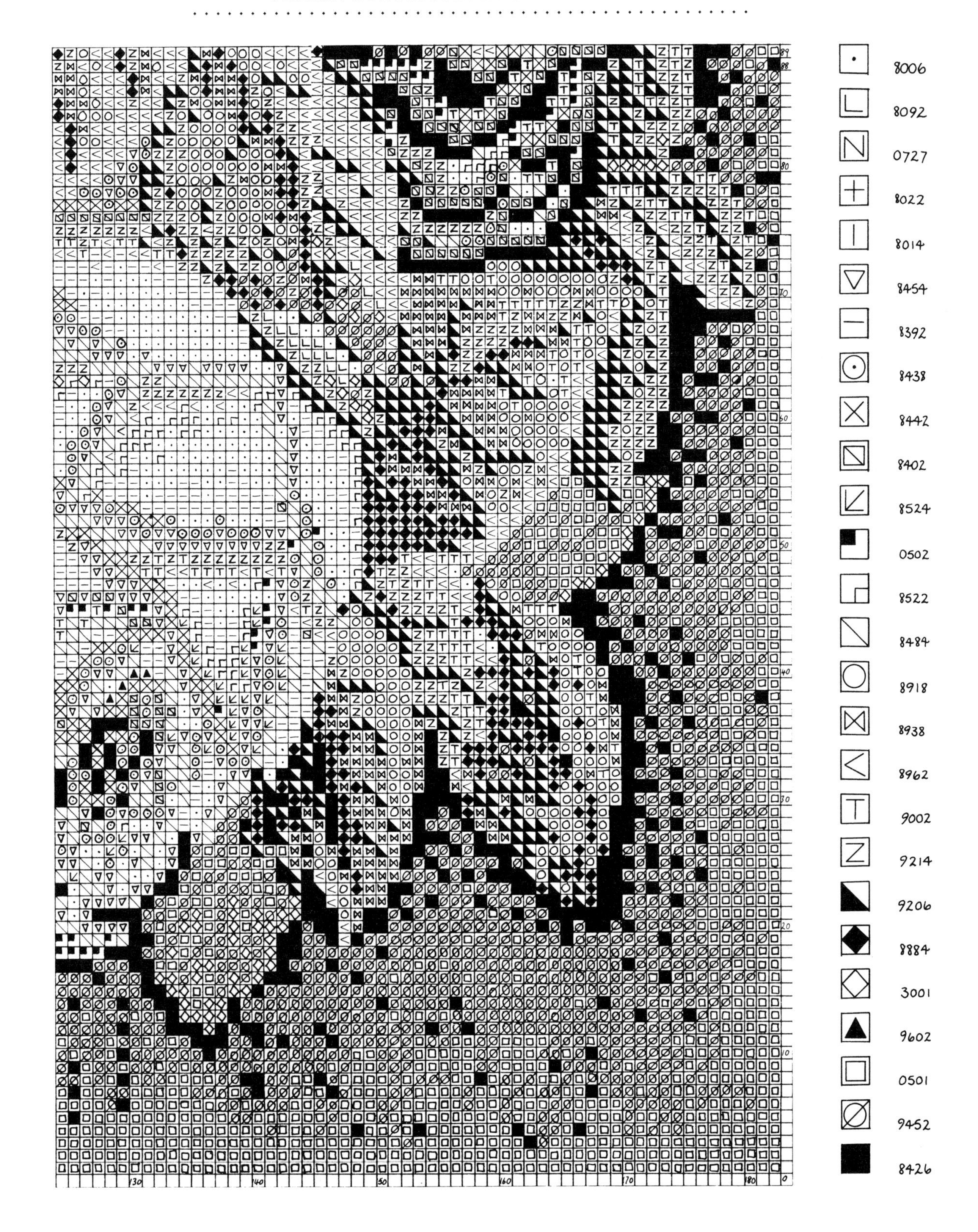
8006
8092
0727
8022
8014
8454
8392
8438
8442
8402
8524
0502
8522
8484
8918
8938
8962
9002
9214
9206
8884
3001
9602
0501
9452
8426

Strawberry Tea Cosy

BY GLORAFILIA

Another lovely and practical piece, this tea cosy combines tent stitch with satin, brick and stem stitches. This needlepoint is suitable for experienced stitchers or the less experienced looking for a challenge.

MEASUREMENTS
Finished size of design 31 × 26.5cm (12 × 10½in)

MATERIALS
White interlock canvas, 12 holes to 2.5cm (1in), 41 × 38cm (16 × 15in)
Appleton's tapestry wool, as follows:

Yarns			*Skeins*
A	881	Cream	3
B	751	Sugar Pink	1
C	221	Dusky Pink	1
D	205	Terracotta	2
E	851	Custard	1
F	902	Ochre	1
G	603	Pale Mauve	1
H	712	Plum	2
I	742	Sky Blue	2
J	323	Air Force Blue	2
K	353	Pale Green	2
L	293	Dark Green	3
M	715	Aubergine	2
N	224	Rose	2

Tapestry needle, size 18
Ruler or tape measure
Masking tape to bind edges
Sharp scissors for cutting the canvas
Embroidery scissors
Sharp HB pencil or fine permanent markers in suitable colour(s)

Preparing Your Canvas
1 Enlarge the chart on p 98 on a photocopier until it measures 30cm (12in) wide along the bottom.
2 Cut the canvas to size, and bind the edges with masking tape.
3 Place the canvas over the enlarged photocopy and trace through with an HB pencil or permanent marker in black. Follow the thick outlines freely, ignoring the canvas grid.
The fine lines are to show colour changes. These can be traced through using a contrasting colour, for example, red, or – if you feel confident enough – it is easier to put them on afterwards free hand using the diagram as your guide.

Stitching the Design
The whole thread of tapestry wool has been used throughout. Wool colour references (letters A–N) and position and direction of the stitches are given on the chart on p. 104. The solid-head arrows show the direction of the stitches. Open-head arrows are used as pointers to show where a line or block of colour is in a particular colour or stitch. Some stitchery has been used: satin stitch (2), stem stitch (3) and brick stitch (4). The remainder is tent stitch (1) – if you prefer, the whole design can be worked in tent stitch. Sometimes only one reference number appears on, for example, a bunch of grapes, but the reference applies to the whole bunch. As a general guide, the central design and background are entirely in tent stitch (1) except for cherry stems and basket handle; where grapes are satin stitch (2), they are outlined in stem stitch (3); any other stems or lines needing emphasis are in stem stitch (3). Some of the fruits and leaves have been worked in vertical or horizontal brick stitch (4) over two threads where indicated. White highlights, on the grapes, are shown in black blocks. Dark green strawberry seeds are

worked randomly and should be sewn before you sew the rest of the strawberry. The basket containing the strawberries is worked in Air Force Blue (J), with random patches of Cream (A), Sky Blue (I) and Custard (E).

Stretching

When the design has been sewn, the needlepoint may have to be stretched back into shape. If it is out of 'square', lightly spray with water and leave for a few minutes to soften the canvas. Gently pull into shape and pin out, right side down, on to blotting paper on a clean flat board. Use tacks, staples or drawing pins and pin outside the sewn surface. Do not strain the canvas too tightly or the needlepoint will dry with a scalloped edge. When the needlepoint is thoroughly dry, remove it from the board. It may take two or three days to dry.

Making Up

Cut a piece of quilted material 1cm ($\frac{1}{2}$in) larger than the finished needlepoint*. Cut the lining the same size as the quilting and place quilting and tapestry right sides together and machine. Turn up bottom and oversew. The inside of the canvas can be lined if you wish. Machine the lining, right sides together, and place inside cosy. Turn bottom of lining up and slip stitch to the bottom of the finished needlepoint. A loop can be added from the same fabric.

* For a professional finish, the tea cosy can be piped before putting on the quilting. Pipe the tapestry, cutting the cord just above the finished work at the bottom of the tapestry.

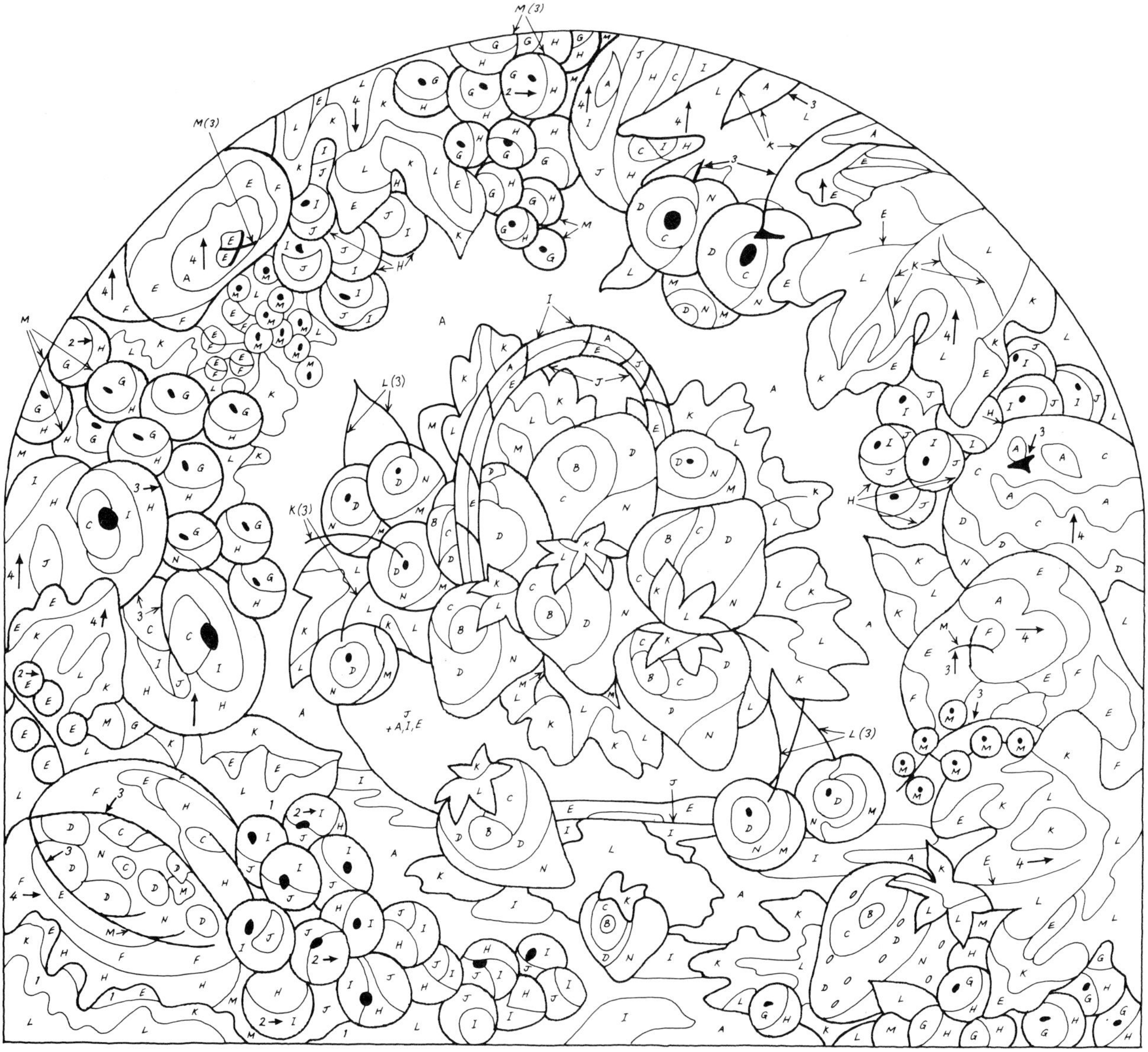

→ Solid-head arrow: stitch direction
→ open-head arrow: pointer

Hooky Rug

BY ANN DAVIES

This hooky rug is a quick and simple way to get started with rug making. The design can be simplified for those wanting a really easy starting piece and the size can be varied according to your needs.

MEASUREMENTS
Finished size 63 × 92cm

MATERIALS
Fine or 2oz hessian (the best quality you can obtain, not upholstery hessian)
Fabrics*
Carpet binding, washed to avoid later shrinkage
Tracing paper, transfer pen or chalk, and waterproof medium and thin felt-tip pens
Staple gun and staples or drawing pins
Scissors or a rotary cutter and cutting mat
Frame (artists' stretchers or four pieces of wood put together firmly), see step 2
Rug hook with barbed point, not a latch hook

Directions
Outline the diameter of your rug by running a fine-tipped waterproof felt-tip pen down between two threads of hessian.

1 First, transfer the design to the hessian. This can be done either by enlarging the design on a photocopier and then tracing it with a transfer pen and ironing the transfer on to the hessian, or by drawing a matching grid pattern on both the sketch and the hessian and carefully copying the outline on to the hessian with chalk, square by square. When you are satisfied with the drawing, go over it with a waterproof medium felt-tip pen. You may find you want to change the design slightly as you hook – perhaps the material you are using suggests variation or you think a shape would look better slightly altered.
2 Stretch the hessian tautly over a simple frame. You need to have a frame at least 10cm (4in) larger on all sides than the finished size of the rug. Begin by stapling or using drawing pins on two sides to make an L-shape, keeping the threads of the hessian straight; the edge of the frame acts as a guide. Then, stretching the hessian tightly, staple along the other two sides, using plenty of staples or drawing pins. Pay particular attention to the corners, bringing the staples right down to the edge.
3 Using either scissors or a rotary cutter and cutting mat, cut the fabrics into strips 5mm–1cm ($\frac{1}{4}$–$\frac{1}{2}$in) wide and approximately 38–45cm (15–18in) long.
4 Working from right to left, the hook is held in the right hand above the hessian in the way you would hold a pencil. Take a strip of material in your left hand and hold it loosely between the forefinger and thumb underneath the hessian. With the curved tip facing upwards, push the hook firmly through the hessian and let the tip scoop up the wool strip. Do ensure you make your hook work for you by pushing it firmly into the hessian to make a large hole so you don't have to struggle to bring up the material. Pull the end up to the right side of the hessian to a height of about 5mm ($\frac{1}{4}$in). (All ends are pulled through to the right side, not hanging underneath, and later cut off level with the other loops. They become invisible in the pile.) Working from right to left, and leaving about two threads between each (more if your strip is very wide) keep pulling up the loops as evenly as you can. The diagram on p. 107 shows the basic hooking.

* I prefer to use woollen fabrics gleaned from jumble sales, car boot sales, ends of roll sales or similar sources. However, pure wool can be difficult to find and synthetics and other materials are often used by rug makers.

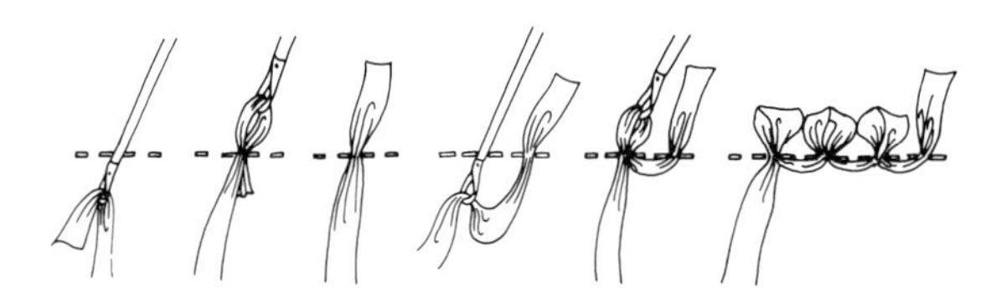

Pull up your loops to a height that appeals to you – not so high that people catch their heels in the pile and not so low that the pile will wear quickly. A height of about 2.5cm (1in) is usual. The surface should be firm but not over packed. Start a new strip in the same hole as the end of the previous strip. Don't feed the material on to the hook, let the hook find the material. Think 'in, scoop, loop'.

5 When making a second row of hooking, begin close to the first row, either above or below it, leaving one or two threads of hessian between. You may at first find you are pulling out your previous loops as you hook. This could be caused by holding the strip too tightly underneath the work and not releasing it when the hook catches the strip to bring it through the hessian. Don't be tempted to run the strip of material from one point of the work to another. If you do so, the hook could snag the material.

Don't work horizontally across your design, work more in a random fashion unless you want definite lines.

6 On completion of the rug, remove it from the frame and cut away surplus hessian to about 4cm ($1\frac{1}{2}$in) all round, cutting across the corners to prevent a build-up of material there. Sew the carpet binding on the right side as close as possible to the last row of hooking. It helps if your binding is a similar colour to that of the last row of hooking. Ease the binding around the corners of the rug; do not allow any excess for mitring.

Turn down the hessian on to the back of the rug and tack it down, catching the hessian lightly to the reverse of the rug. Turn the binding down over the hessian and hem it down, again catching the stitches in the reverse of the rug. Coax the excess binding at each corner into a mitre and catch it down. You can sew over the mitred corners if you wish. Hem the binding all round the rug, ensuring that it covers the turned down hessian. A useful tip is to place a few strips of the fabric between the hessian and the binding so that if, at a later stage, there are any disasters, then you have some fabric to do running repairs.

Place the rug right-side down on a towel or blanket and, covering it with a damp cloth, press all over with an iron.

Victorian Tile Patchwork Block (Cushion)

BY FREDA PARKER

This is a very simple but effective patchwork cushion which can be stitched either by hand or machine. It's a lovely, quick project for all patchwork fans.

The origins of patchwork are lost in the mists of antiquity – fragments have even been found in the tombs of ancient Egyptians. Very little has survived, however, from before the eighteenth century, but there is a wealth of nineteenth-century work still to be seen in museums and private houses. During the early part of the twentieth century interest in patchwork faded and at one time it seemed the craft would be lost. However, over the last twenty or thirty years there has been a steady revival, with patchwork becoming probably one of the most popular pastimes.

There are basically two different types of patchwork, usually thought of as the 'English' and the 'American' methods. With the first, small, geometric shapes which have previously been tacked to thin card are oversewn together. The most well-known shape for this type of patchwork is the hexagon.

With the second type, the patches are seamed together with running stitches or with a straight machine stitch. Shapes are usually simple squares, rectangles and triangles. This method has always been the most commonly used in America, but it was also very popular in the north of England, Wales and Ireland. The technique is used for making medallion quilts in which a central design is surrounded by a series of borders; for strip patchwork, where lengthwise strips made up of squares and triangles are alternated with plain strips; and for making block patterns.

There are numerous ways of arranging squares, triangles and rectangles to make up a square block which can be used to create a repeat design. This technique was particularly popular with early settlers in America. It was simple to work with lap-sized pieces and a number of people would piece blocks for the same quilt. There are hundreds of different traditional block patterns, some of which have romantic and evocative names – Columbian Star, Philadelphia Pavement, Virginia Reel and so on.

The design for this cushion is based on part of the Victorian tiled floor in the hall at Highclere Castle, near Newbury in Berkshire. The block can be used on its own to make a cushion or repeated to create a wall hanging or quilt. The finished size of the block shown opposite is 32cm (12½in) square, but the template can be drawn to any size you choose.

MEASUREMENTS

Finished size of block 32cm (12½in) square

MATERIALS

Closely-woven cotton fabric in four different plain colours or small all-over patterns
Cotton thread to match fabric*
Squared paper
Pencils (lead pencil for making template and blue or yellow pencil for marking fabric)
Medium-weight cardboard and paste
Metal ruler
Cutting board and craft knife
Sandpaper
Dressmaking scissors for cutting fabric
Small scissors for snipping off threads
Fine dressmaking pins

* Match thread to work wherever possible. If sewing by hand you can change the colour to suit the patches being joined. When joining a lighter patch to a darker one, choose the darker colour, as dark on light is less obvious than light on dark. If using a machine, choose a colour which predominates in the work.

Making Up

You will need to add a 6mm ($\frac{1}{4}$in) seam allowance all around each pattern piece when cutting out.

1 Draw out the pattern to the required size on squared paper. Paste this pattern on to the card. It is a good idea at this stage to note the colour on each section of the pattern.

2 Using the metal ruler, craft knife and cutting board, cut out the pieces of card accurately. Use the sandpaper to smooth any rough edges.

3 Following the pattern and using a well-sharpened pencil in a colour which will show up on the fabric, place each pattern piece on the wrong side of the appropriate fabric and draw round it accurately. Make sure that the squares and rectangles are on the straight grain of the fabric and that the triangles have their two short sides on the straight grain. This is essential to ensure a good shape when making up the block.

4 Cut out the fabric 5mm ($\frac{1}{4}$in) from the drawn line (this is the seam allowance).

5 Pin two patches with right sides facing and sew together by hand or machine, from raw edge to raw edge, along one sewing line. Accuracy is essential or the pieces will not fit together properly. Start by sewing rectangular patches to two opposite sides of the central square (thus making a strip). Press seams either to one side or open. Next, sew a small square to each end of each of the other two rectangles and press. Join the strips just made to either side of the central strip and press. You now have a square. Make up all four corner sections in the same way. First join two triangles in contrasting colours along the long sides to make a square. Press, then add the two outer triangles. Press. Take two of these large triangles and sew one to each of two opposite sides of the square. Press. Repeat with the other two large triangles. Press.

Make up into a cushion.

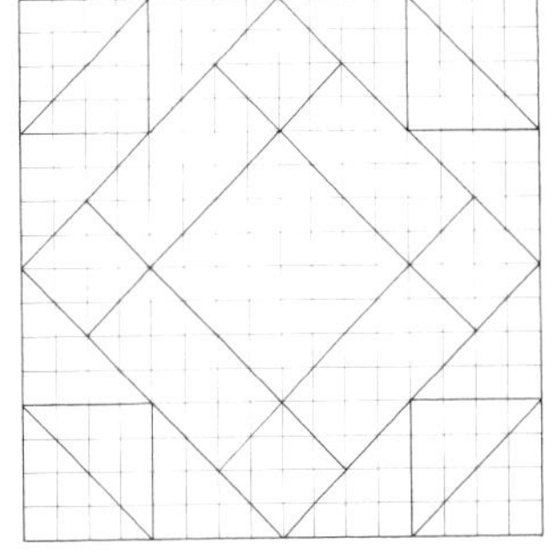

Appliquéd Drawstring Bag

BY JANE WALMSLEY

This is a truly stunning appliquéd bag. It is hand-stitched and requires a high level of competence, but is well worth the time involved.

MEASUREMENTS

Finished size 20cm (8in) square

MATERIALS

Use finely-woven 100% cotton dressweight fabrics throughout
25cm (10in) of 90cm wide fabric in cream and grey-mauve
25cm (10in) of 90cm wide lightweight sew-in Vilene
15cm (6in) squares of bright mauve, dark mauve, pale gold, green-gold and mid-green fabrics.
5cm (2in) squares of fabric in terracotta and peacock blue
Sewing threads to match appliqué fabrics
Stranded embroidery threads in three shades of: green, of tan and also mid-grey
3m (3 yards) of gold cord
Very small amount of polyester filling
Long, thick darning needle
Sheet of tracing paper and thin white card for templates
Light blue colouring pencil
Fine, permanent ink, black fibre-tip pen

1 Cut out two pieces of cream fabric 22.5cm (9in) square (this includes 12mm/$\frac{1}{2}$in seam allowances) for the bag front and back, two pieces of grey-mauve fabric for the lining and two pieces of Vilene for the interlining, the same size as the cream fabric.
2 Using the fibre-tip pen, trace the complete design on to the tracing paper (see chart on p. 116 and photocopy to double the size – 200%, tape the traced design to a flat surface and slip a sheet of white paper underneath. This will show up the traced design more clearly. Centre a square of cream fabric, right side up, over the traced design and tape in position. Draw the design on the cream fabric with the light blue colouring pencil. Ensure that the pencil point is kept sharp so the traced lines are fine and clear.
3 Make several round card templates for the grapes, one large and one small leaf template (see chart on p. 116) and the five templates for the butterfly (see Figs 1 and 2 and increase both by twice the size they are shown – 200%). Place the traced design over the white card and, with a hard pencil, draw around each of the shapes. Lift the tracing and indentations will be seen on the card. Cut out the shapes carefully on these lines and mark the dotted lines.
4 To make the stems, cut out three strips, 2.5cm (1in) wide, of pale gold fabric on the true bias. Following the numbered sequence shown on the chart on p. 116, stitch stems in position thus: fold the bias strip in half lengthways with the folded edge of the fabric on the inside of the curve, pin the strip to the inside stem line (see Fig. 3), stitch in place with a running stitch strengthened with an occasional backstitch (see Fig. 4). It will be necessary to snip the cut edges to allow the strip to be curved around the marked shape. The stem ends will be covered by subsequent stems, leaves, grapes, etc.
Trim excess fabric (Fig. 5), then roll folded edge over seam allowance, stitch in position (Fig. 6). Following the numbered sequence, complete all the stems in this way.
5 To make the grapes, using the circle template (Fig. 2) cut four grey-mauve, five bright mauve and four dark mauve circles of fabric. Using matching thread, sew a gathering thread around a bright mauve circle 3mm ($\frac{1}{8}$in) from cut edge (Fig 7).

Place a card circle on the wrong side of the fabric and pull the gathering thread up tight around it (but do not secure the thread). Keeping the thread taut, iron the circle on the right side with the card still inside (Fig. 8).
Loosen the thread and remove the card circle. Place a little filling inside the fabric circle and gently pull the gathering threads so that a slightly domed shape is formed. Secure the thread with a backstitch (Fig. 9).
Align the padded grape, gathered side down, on the traced design on the bag front, covering the end of the stem; stitch in position. Make all the grapes in the same way.
6 To make the leaves, trace the large leaf on chart (p. 116) and draw around it on the wrong side of the green-gold fabric. Cut out, adding 6mm ($\frac{1}{4}$in) turnings.
Place the fabric leaf shape on a folded piece of fabric, wrong side up; place template (reversed) on top. Hold the template firmly in place with one hand; hold the darning needle in the other hand and, keeping it almost horizontal with the fabric, press the needle firmly all round the template. (It is usually easier to stand up to mark the fabric.)
Remove the template and turn the leaf shape to the right side, a sharp crease outlining the leaf shape will be seen.
From the right side, turn the seam allowance to the wrong side and tack as carefully as possible. On the leaf points turn the first side under and press between the finger and thumb, then turn the other side under (see Fig. 10); on the sharper points this will leave a 'tail'. Do not cut this off and do *not* iron the leaf at this stage.
Tack the leaf in position on the marked design covering the stem end. Use small hemming stitches, angle the needle so that it comes out through the fold, having picked up some of the seam allowance underneath, so making the stitches almost invisible. Always keep the thumb firmly on the fabric being appliquéd. Stroke the 'tails' on the leaf points in and mould the fabric with the needle as sewing progresses. Snip into the inner curves; using the needle point, fold fabric under, but at the base of the inner point on raw edge, strengthen with a neat semi-circle of

small stab stitches over the edge.
Make the small leaf template as for the large leaf template and cut one from the mid-green fabric and, reversing the template, one from the green-gold fabric. Mark, tack and stitch as above.
7 To make the butterfly, reverse the templates (Fig. 1) and cut one shape from each as follows, adding 6mm ($\frac{1}{4}$in) seam allowances: B1 in terracotta, B2 in grey-mauve, B3 in peacock, B4 and B5 in grey-mauve. Needle mark around the shapes as described above but tack the seam allowances back along the solid lines only.
B1: align the folded edge on traced butterfly (Fig. 11) and stitch in position.
B2: position and stitch prepared shape B2, which will partly overlap the cut edge of B1 (Fig. 11).
B3: position and stitch as indicated on Fig. 11).
B4: abdomen – position and stitch in place.
B5: head – stitch in place.
Thorax: using the grape template (Fig. 2) cut 1 circle from the grey-mauve fabric *without* seam allowances. This time, however, run a gathering thread all round circle 3mm ($\frac{1}{8}$in) in from cut

edge. Place a small amount of filling on the wrong side of the circle; pull gathering thread up tightly and secure. Position between head and abdomen, raw edges underneath, and stitch in position, making first four stitches at the top, base and sides. Keep stitching round and round invisibly until a small, tight, half sphere has been formed.

8 For the embroidery use a single strand of embroidery thread throughout. See Figs 12, 13, 14 and 15 for stitches.

Tendrils: work in stem stitch, sewing two rows in two shades of green very close together.

Leaf veins and butterfly antennae and legs: work single rows of stem stitch in appropriate colours.

Butterfly wings: work straight lines of split stitch in grey embroidery thread over B3.

Work chain stitch circles and scalloped line in tan.

9 To make up, use 12mm ($\frac{1}{2}$in) seam allowances. Tack Vilene to the wrong side of the bag back and front. Place lining over bag front and bag back, right sides together. Pin and stitch together along the top edges. Press the seams open (Fig. 16). Place these two sections right sides together, bag front to bag back, lining to lining. Stitch around outer edges, leaving an opening at the base of the lining and 12mm ($\frac{1}{2}$in) openings on both sides of the bag front (Fig. 17). Snip seam allowances on curves almost to stitching lines. Turn through to right side through opening in lining. Stitch opening together.

Tuck lining inside bag, tack along seam line and make a further row of tacking 12mm ($\frac{1}{2}$in) down. Work buttonhole stitch by hand or machine around the four slits on the bag front and back as indicated.

Stitch the cord around the bag, making a loop and tucking the cord ends into the openings left either side of the bag top, stitch these openings together. Cut two pieces of cord 75cm (30in) long. Thread one length of cord all round from the right and the other length all round from the left. Bind each pair of cords together 4cm ($1\frac{1}{2}$in) in from the cut ends. Unravel the cord ends to make tassels (Fig. 18). Cut a 12cm ($4\frac{3}{4}$in) length of cord; thread through the loop at the base of the bag; bind cords together, and fray to make a tassel.

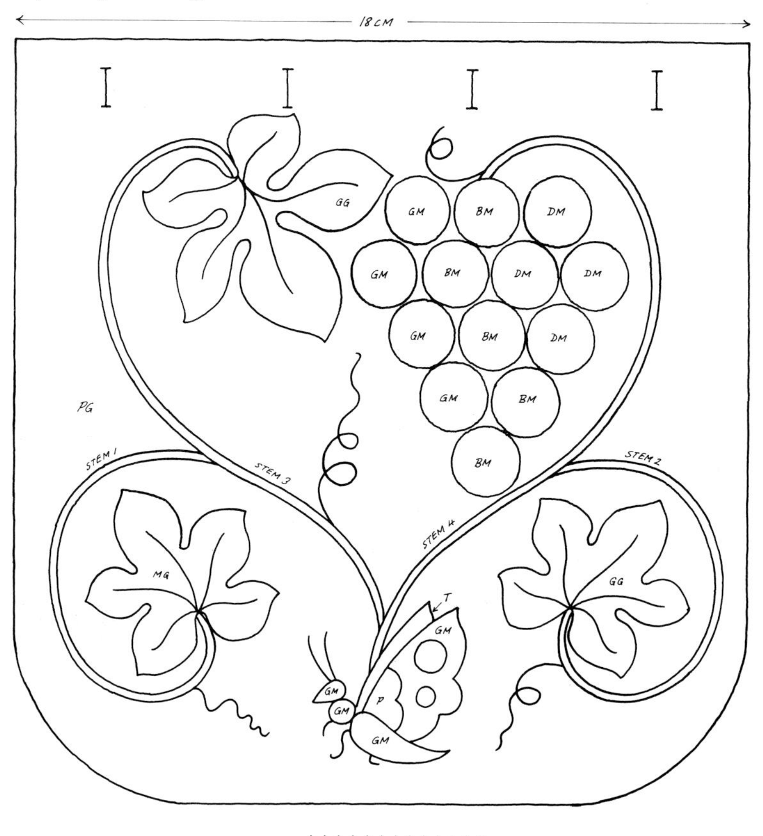

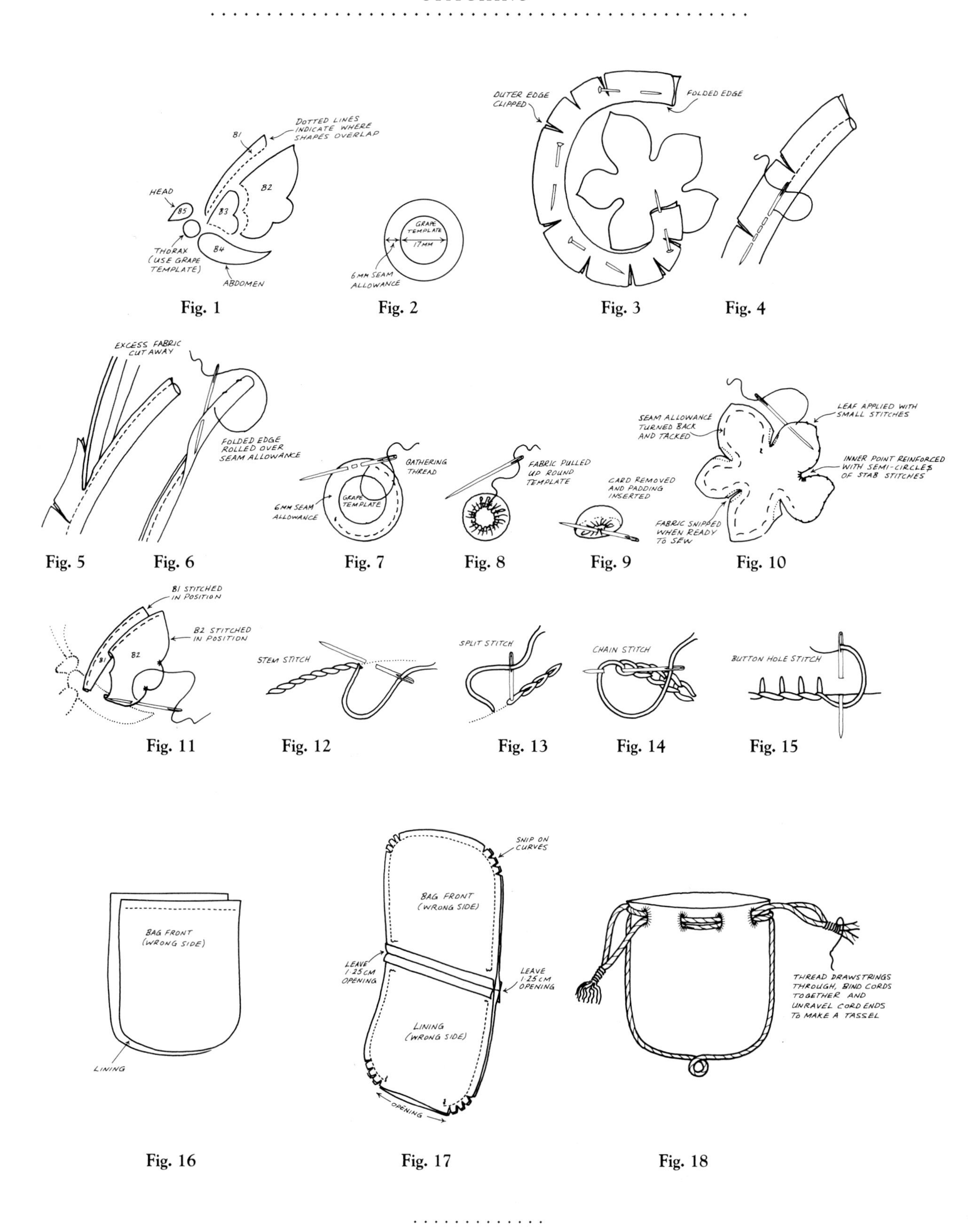

Fig. 1 Fig. 2 Fig. 3 Fig. 4

Fig. 5 Fig. 6 Fig. 7 Fig. 8 Fig. 9 Fig. 10

Fig. 11 Fig. 12 Fig. 13 Fig. 14 Fig. 15

Fig. 16 Fig. 17 Fig. 18

A Small Treasure Bag

BY GAIL HARKER

This treasures or jewellery bag is made using machined patchwork and makes an exciting challenge for any experienced sewer. It is a beautiful item which would make an ideal gift – if you could bear to give it away!

MEASUREMENTS

Finished size 15 × 14.5cm (6 × 5¾in)

MATERIALS

Patchwork – 4 pieces of printed cotton fabric for squares. All fabric should be of the same weight. Choose 3 colours that are similar in colour and pattern. The fourth could be a different shade. Medium pink, hot pink, peach and yellow with patterns of various colours have been used here. A 25cm (10in) square of each colour will allow you to cut a few extra squares. You may then move these about to arrive at the most pleasing arrangement.

Lining fabric – 19 × 36cm (7½ × 14in), includes seam and turning allowances; choose a plain colour in a complementary colour to the patchwork squares.

Flap Lining – 11.5 × 16.5cm (4½ × 6½in), includes seam allowances; choose a lightweight fabric flattering the other colours.

Binding – for the front flap and inside edge of the bag. Either use manufactured binding or make your own: one strip 16.5 × 2.5cm (6½ × 1in) and another strip 16.5 × 5cm (6½ × 2in). Select colours to complement the lining material.

Wadding – use a 20 × 40cm (8 × 16in) piece of 2oz wadding or substitute flannel or any fabric with a bit of bulk, such as pre-shrunk blanket material.

Optional – if your machine feed does not like stitching on the wadding, use a piece of cotton muslin or other thin fabric as backing under the wadding.

Threads – to decorate the squares, use colours of machine embroidery or dressmaking threads in colour tones deeper than the fabric colours of the squares. To stitch the bag together, use No. 50 dressmaking cotton or synthetic thread.

Fastening – decorative button and 10cm (4in) strand of braided cotton thread to match bag colour (see Closure).

Patchwork Instructions

1 Trace the template from light card measuring 6cm (2½in) square. This provides a seam allowance of 5mm (¼in) around the square. You will need 21 squares. Allow a seam line of ½cm (¼in).

2 Press the fabrics. Lay the template on the fabric and trace around using a hard pencil. Try to align the edges of the template with the straight grain of the fabric (parallel to the selvedge).

3 Cut out the squares. The bag will require 21 squares to be cut from the 4 colours of patchwork fabrics. If all the patchwork fabric is cut into squares, there will be more than the required number.

4 Lay out the squares three wide by seven long in different arrangements until you find the arrangement you like the best. Use the extra squares to find an arrangement of colour that suits the eye.

5 Stitch the squares together by machine using a 2.5mm (⅛in) stitch length (see Fig. 1 on p. 117). First, join the rows of three squares. After all the strips of three squares have been joined, press the seams to one side. Alternate sides as you press them. This will make for a strong seam and make it easier to line-up seams as the strips are joined.

6 Lay the front sides of two of the three-square strips together and carefully pin them together,

aligning the seams as you go (see Fig. 2 on p. 117). Stitch these together using the 5mm ($\frac{1}{4}$in) seam allowance. Continue until all seven rows of three-square strips have been joined. Press the seams to one side when finished. You will now have the patchwork ready for the next process.

Preparations for Quilting

1 Place the wadding over a piece of cotton muslin. Over this, lay your completed patchwork (right side up) leaving a 2.5cm (1in) margin of wadding and muslin all around. The margin gives an extra edge to hold while you are doing the decorative stitching. This makes a three-layer sandwich of fabric.

2 Starting from the centre, securely baste all the layers together (see Fig. 3 on p. 117).

3 Using synthetic or cotton threads to match patchwork fabrics, stitch in the seam across the three-square strips using sink stitch, with a 2.5mm ($\frac{1}{8}$in) stitch length: holding the fabric on each side of the seam, gently stretch as you sew (do not stretch too much or squares will part). This will allow the stitches to almost disappear between the seams. After sewing across all the three-square strips, turn the fabric and, using the same technique, stitch down the length of the fabric following each seam.

Decorating the Patchwork

1 Thread your machine with the threads you have chosen to decorate the fabric. Start on the square in the centre of the patchwork (see Fig. 4, p. 117). Begin at any edge, working from the seamline of the square. Apply a few backstitches at start and finish of stitching in each square to secure threads. Alternatively, leave long threads to take to the back of the square and tie off by hand. As you approach within 5mm ($\frac{1}{4}$in) of the adjacent edge, stop stitching, leaving the needle in the fabric and, using the needle as a pivot, turn the fabric to follow the next edge. Do the same at each corner. Continue until you have completed several rows of stitching, finishing in the centre of the square. Let the sewing foot act as a guide in stitching lines the same distance apart.

2 Working from the centre square, continue stitching the pattern into each square. Do not be concerned if the spiral patterns are not identical. As an alternative, the spirals could be stitched using a narrow zigzag stitch.

3 Trim the excess muslin and wadding up to the edge of the patchwork.

Joining the Lining to the Patchwork

1 Press the lining. Place the lining with the *wrong side* facing you. Along both of the long sides of the lining, turn over 5mm ($\frac{1}{4}$in) and press down.

2 Position the patchwork, *finished side up*, on the lining so that the turned edges of the lining overlap the edges of the patchwork by 5mm ($\frac{1}{4}$in).

Make a second turn of the previously folded edge of lining, folding it over the patchwork, and press. This will create a fold of fabric over the edge of the patchwork. Pin this in place.
3 Where the edge of the fold on the lining meets the patchwork, apply slip stitch along both edges of the lining, by hand, joining the lining to the patchwork (see Fig. 5, p. 117). The stitching could be done by machine, using straight stitch, but this may flatten the edges slightly.

Binding Ends

Bind the ends of the bag, using the two strips described in the materials list. The binding for the front flap is the 2.5 × 16.5cm (1 × 6½in) strip.
1 Fold over 5mm (¼in) along a long edge and the two ends and press.
2 Pin the right side of the strip (raw edge) to the lining side of the patchwork. Straight stitch by machine.
3 Cut away the excess wadding and seam allowance and press all seams to one side.
4 Fold the binding over the end and to the front of the patchwork, taking 5mm (¼in) of the lining with the binding.
5 Roll 5mm (¼in) of binding under and pin to the patchwork. This will create a two colour strip composed of 5mm (¼in) of lining material and 5mm (¼in) of binding.
6 Slipstitch the folded edge of the binding to the patchwork. *The bottom edge is bound with the remaining strip of fabric*, 16.5 × 6cm (6½ × 2½in).
1 As with the top, fold over and press down 5mm (¼in) across the long edge and both ends of strip.
2 Pin the raw edge of this strip, *right side to right side*, to the raw edge on the *front* side of the patchwork, and straight stitch by machine.
3 Press the seam to one side and cut away excess wadding and seam allowance.
4 Fold the binding over the end of the bag to the underside (the lining side) of the patchwork, and slipstitch to the lining.
N.B. These two strips are sewn on opposite sides of the patchwork.

Sewing the Bag Together

1 Measure 12cm (4¾in) from the bottom edge of the bag and fold along this line (see Fig. 6, p. 117). This will give you an envelope shape.
2 Pin the edges together and, using overcast stitch, sew the edges together with dressmaker's thread. Work on the finished side of the bag.

Flap Lining

1 Turn under 5mm (¼in) along all the edges of the flap lining and press. This should allow approximately 5mm (¼in) of the bag lining as a border around the flap lining.
2 Pin in place and slipstitch to the bag lining around all sides (see Fig. 7, p. 117). Leave a small gap for the bottom loop (see next instruction).

Closure

Before final stitching of the top edge of the flap lining, make a small braided button loop. Plait three lengths of six-stranded cotton or Perlé thread and stitch over the ends to stop unravelling. Slip the raw edges under the flap lining. Pin into position and stitch using a few inconspicuous stitches before stitching down the flap lining. Make sure the loop is large enough to allow the button to pass through.

Tips for Making Up the Bag

Overcast stitch is used to stitch the bag together at the binding. Working from the *finished side* of the bag, begin stitching a short distance from the corners of the bag. Stitch back into the corner and then back up the side of the back. The double stitching will help to reinforce the corners of the bag. Make very small stitches, catching a few threads of the binding with each stitch. When the end is reached, return a few stitches back down the bag as for the corners. This will help to make a strong corner. Use dressmaking threads for the stitching, using a thread colour that is the same as the binding, and try to make approximately 15–20 stitches per 2.5cm (1in).
Slip stitch should be almost invisible. This stitch is used to sew bindings and linings to the patchwork. With the threaded needle, pick up as small an amount of fabric as possible from the background. Slip the needle into the turned edge of the lining. Again, going to the background fabric, pick up a little fabric and continue until completed.

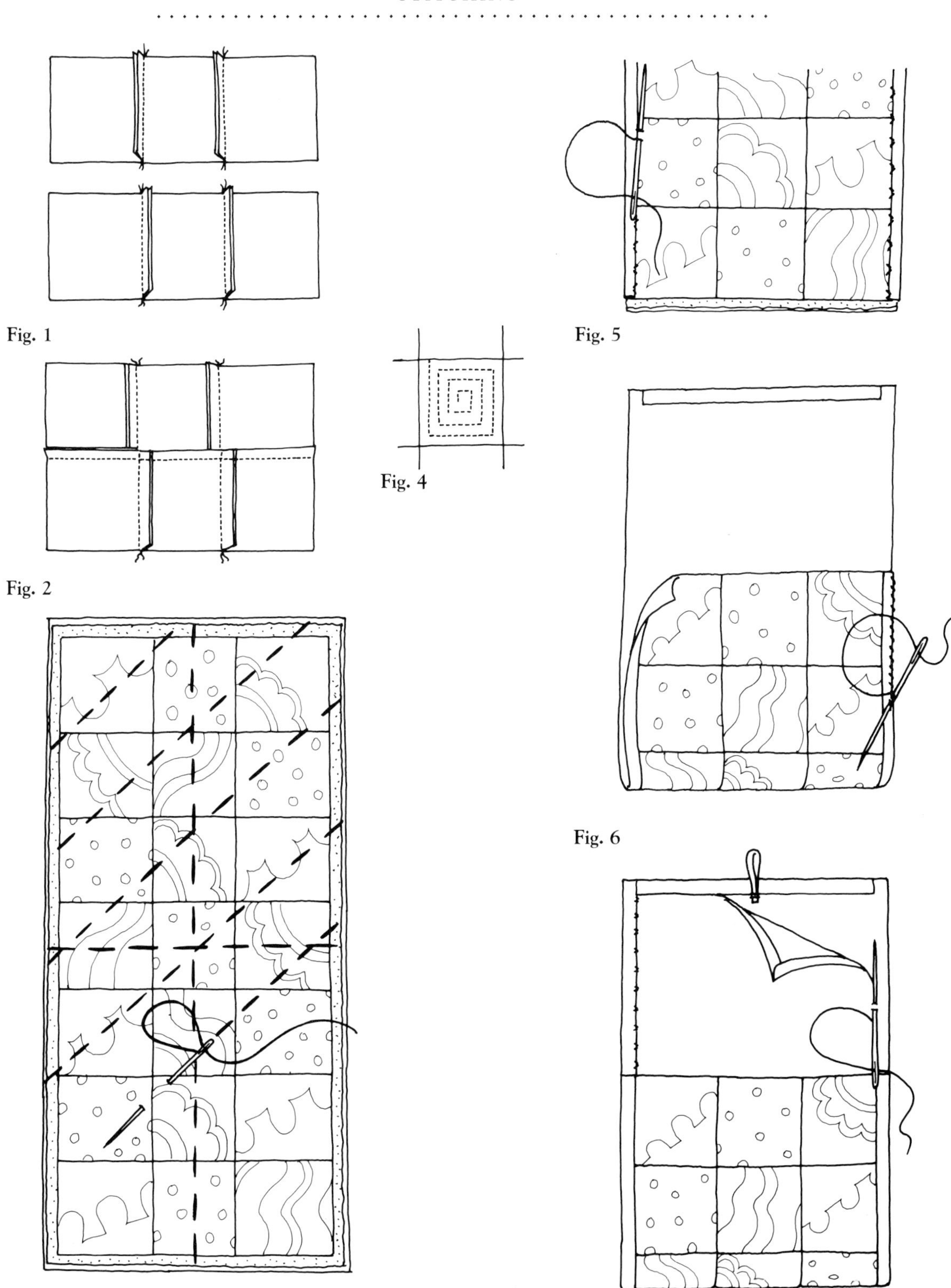

Fig. 1

Fig. 2

Fig. 3

Fig. 4

Fig. 5

Fig. 6

Fig. 7

Fabric Picture

BY ARIELLA GREEN

This is a highly unusual fabric collage which takes time, patience and a sewing machine able to do free embroidery, but is well worth the effort!

The design for this collage picture is created as you go along, although it is possible to work from a previously prepared design if you wish.

MATERIALS

Cotton/polyester fibre, in white, 50cm (20in) square or larger
Vilene heavy-sew, 50 × 50cm (20 × 20in)
Pure polyester fabric, in white, approximately 20cm (8in) square
Coloured sewing threads
Plain white cotton, approx. 20cm (8in) square
Layout paper (thin white paper) 20cm (8in) square
Oddments of printed medium-weight fabric (optional)
Fabric paint, water jars and brushes
Fabric transfer paint by Deka
Felt-tip pens for fabric
Wax crayons
Slick writers by Inscribe
Dressmaking pins
Scissors
Masking tape

METHOD

The first step is to decide on a general colour scheme for your picture. This will guide you in preparing the selection of colours and shapes that you will use to make the appliqué. If you wish to work from a prepared design, draw it on to paper at this stage. When the design is ready, trace it on to the sheet of Vilene.

MAKING THE APPLIQUÉ PIECES

Most appliqué is made from 'found materials' (i.e. materials left over from other projects, or found by chance) which are then cut up and pieced together. The method of appliqué used here is slightly different in that most of the material to be used is not 'found' but painted by yourself. First, create fabrics decorated with rough shapes and colours appropriate in mood to the final picture, and then cut these fabrics into the pieces for use in the appliqué. The advantage of this method is that you have much greater control of the final outcome, and more possibilities for creativity. These fabrics are made in two ways; firstly, using direct painting on to material, and secondly, using a method of transfer dyeing. Each method gives a slightly different texture and colour to the fabric and adds variety to the final appliqué.

1 Direct Painting

Stretch the white cotton sheet on a hard surface using the masking tape (protect the underneath against paint soaking through). Paint directly on the sheet with the fabric paint, using colours and shapes which are in keeping with your design idea. More detail can be added using the felt-tip pens and wax crayons. The purpose at this stage is not to do a realistic image or 'painting', but to make an abstract colour harmony for later use. As you get experienced at this technique you may find that the preparation of fabric like this allows you to be very free with the feeling and expression which underlie your creation, without having to worry about the final image or form. When the sheet is painted, leave to dry, and then fix the colours by ironing over the back.

2 Transfer Dyeing

Using the transfer paints, paint realistic images of your favourite flowers, leaves and stems on the layout paper. Remember that whatever you paint will become its mirror image after the transfer

printing. Allow the paint to dry. Lay the paper flat (painted side down) on the white polyester and cover with a cotton cloth. Press hard with a hot iron to transfer the design, referring to the instructions on the paint packaging. Lift off the paper and put aside. The image you painted will be printed into the polyester.

You can make as many different fabrics as you like using either of the above methods; they can all be used in the final appliqué.

Building the Appliqué
Place the Vilene on a flat surface. The Vilene is going to be the backing on to which you pin and then stitch the appliqué. With sharp scissors cut, or piece, the painted or printed fabrics into appropriate shapes. You will need to experiment with this to find the size that suits the design, some pieces will need to be large, some small. Build up the image of the vase itself first, arranging the fabric pieces on the Vilene. The process is rather like a jigsaw; the pieces can be moved around and rearranged until you find the right composition. If you find that you are missing some colours or shapes, use the found materials and the felt-tips and crayons as well.
When the vase is finished, cut out flowers and stems from the fabric you have prepared. Build up the arrangements of flowers in the same way. When you feel that pieces are in the right place, pin them to the Vilene with dressmaking pins. When you are happy that the image is complete, and it is all pinned in place, you can begin to sew.

Sewing the Appliqué
Choose a coloured cotton to match the colours of your fabric (you might have to change the colour of sewing cotton a few times but it is worth the effort for a clearer effect). Use the satin stitch embroidery setting on your machine. As you progress, take the pins out. If you are an experienced machinist, you can sew without the foot or hoop as the Vilene is firm enough to hold on its own. If you do take the foot off, follow the machine's instructions for darning or monograms. Hold the work tight and as close to the needle as safety allows (watch your fingers!). When you have finished sewing, cut around the work with sharp pointed scissors. The Vilene stays on the back of the piece and must not be separated from it.
As a final touch you can add beads or a few hand-sewn French knots and designs using slick writers. Mount the work on a card, like a photograph. You can also frame and glaze it if you wish.

Velvet Embroidered Collar

BY PADDY KILLER

This beautiful velvet collar is really only suitable for those experienced stitchers who have the confidence to use their sewing machines for free-embroidery. The final collar is very versatile and will suit many designs.

MATERIALS

Cotton dress velvet, 40cm square
Polyester wadding, 40cm square
Cotton organdie, 40cm square
Lining fabric, to match the velvet, 40cm square
Machine embroidery thread, preferably Madeira Sticku, in self-colour
80's machine needle
Sewing machine capable of free-machining
Darning foot (available separately)
Fur hook

Directions

1 Scale up design (see opposite) on paper, or enlarge on a photocopier. Each square is 1cm.
2 Trace the design onto the organdie with a felt-tip pen. Do not cut out at this stage.
3 Pin and tack the three layers of fabric together; velvet with pile downwards, the wadding in the middle, and the organdie on the top.
4 Prepare the machine for free-embroidery. Using a straight stitch, outline the design.
5 Turn the work over so velvet is on top, and fill in between the shapes, using the machine embroidery technique of 'vermicelli', in other words small closely-worked stitches in straight stitch.
6 Change to normal sewing mode, with a zigzag foot, and satin-stitch along the indicated lines.
7 With right sides together, line the collar. Pin and tack before sewing along the seam line to the dot, leaving the neckline open.
8 Cut along the cutting line, snip the curves, and turn the collar right-side out.
9 Turn in the neckline and slipstitch. Sew on the fur hook at the neckline for a closing.

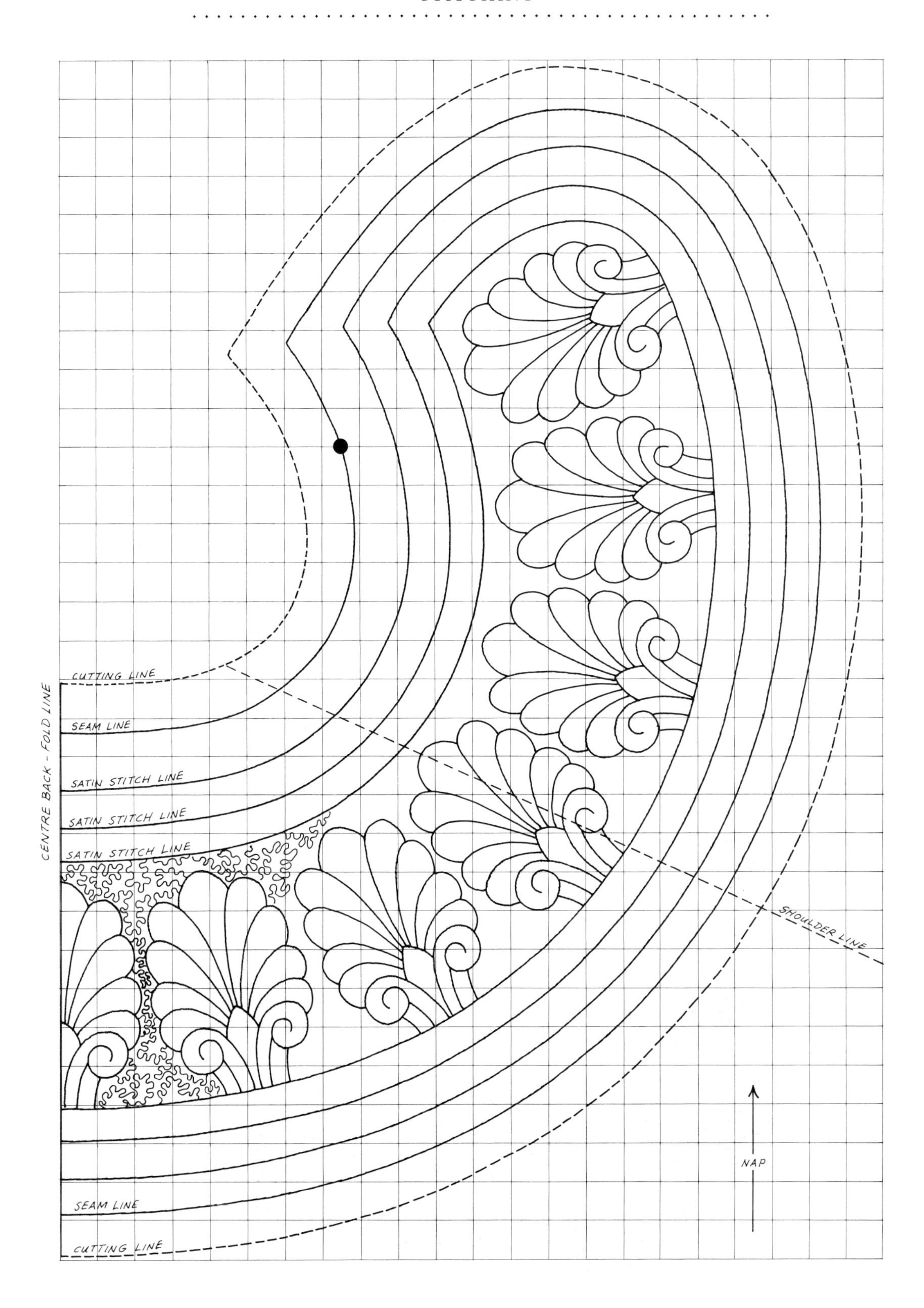
CUTTING LINE
SEAM LINE
SATIN STITCH LINE
SATIN STITCH LINE
SATIN STITCH LINE
CENTRE BACK - FOLD LINE
SHOULDER LINE
NAP
SEAM LINE
CUTTING LINE

Daisy Cushion

BY LOIS VICKERS

This simple but very effective cushion has been made with a variety of stitches including lazy daisy stitch, whipped running stitch and French knots. This is an ideal project if you've never done any embroidery before.

MEASUREMENTS
Finished size 46cm (18in) square

MATERIALS
Evenweave fabric, 24 holes to 2.5cm (1in), two pieces each measuring 48cm (19in) square
A 41cm (16in) zip
A cushion pad, 46cm (18in) square.
DMC Coton Perlé, one skein in each of the following colours – 780, 921, 210, 725, 349; two skeins in colours 783 and 402
Tapestry and crewel needles

METHOD
Cut out the fabric on the straight grain and zig-zag the edges. Stretching in a frame is not essential, but take care with your tension.
Using sewing thread and a tapestry needle to mark the fabric with running stitch, divide the fabric into four equal squares and add two diagonal lines crossing at the centre. These lines are XX and XY on the diagram, which is rotated around the central X point to give the complete design.
Transfer the design onto the fabric within your

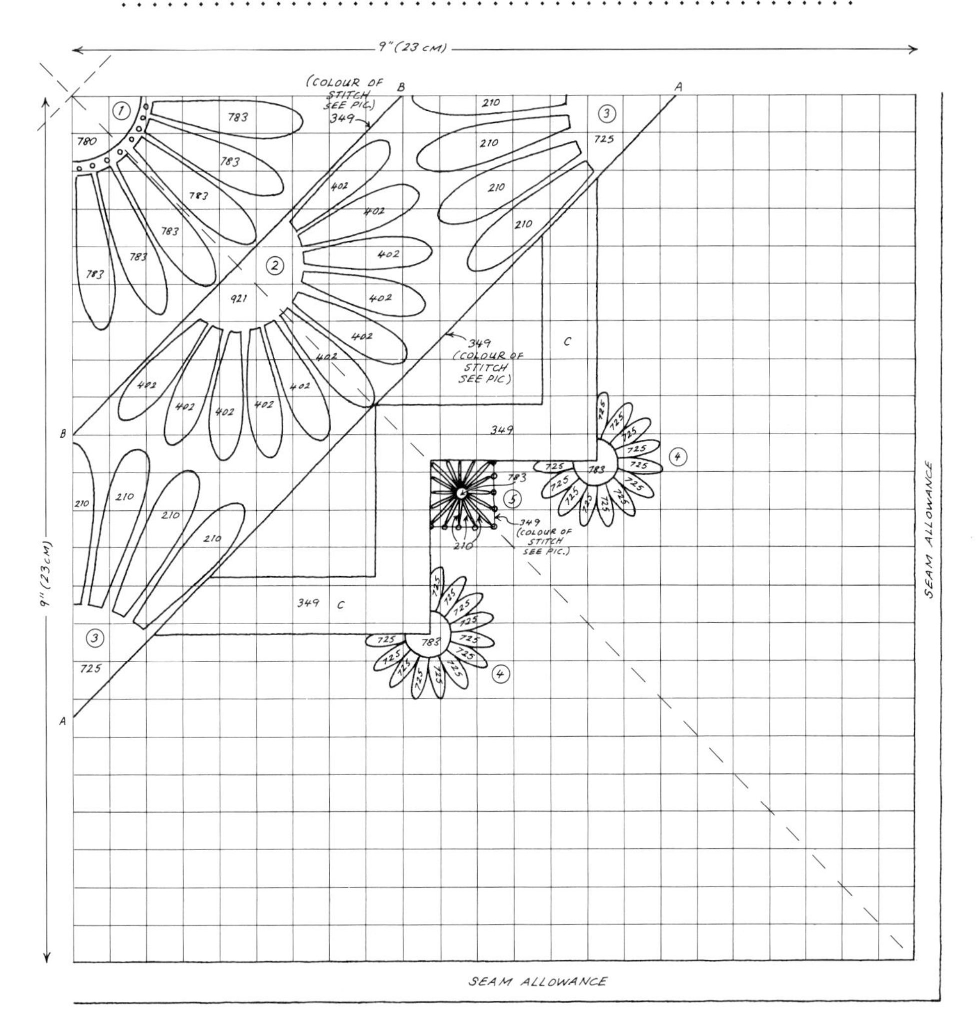

guidelines. Use one thickness of thread at all times in a crewel needle except when stated. Work lines BB to form the central diamond. Use 349 and a tapestry needle, and work small running stitches over and under one pair of crossed threads of the fabric. Whip the running stitch, then weave in and out of the whipping to give a ric-rac effect. Line AA is worked in the same way except that the running stitch is merely whipped not interlaced. Band CC is worked by darning over and under two threads at a time and staggering the rows of darning to give a brick pattern. Use a tapestry needle. Flower 1 has petals in 783, flower 2 has petals in 402, and flower 3 has petals in 210, all worked in Cretan stitch. The flower centres of 1, 2 and 3 are outlined in stem stitch (a double row for flower 1) and then worked in backstitch trellis. Flower 1 centre also has a row of French knots in 780 between the two rows of stem stitch (see diagram). Work flower 1 centre in 780 over a grid of two threads at a time. Work flower 2 centre in 921 over a grid of four threads at a time with a French knot in the centre of each square. Work flower 3 centre in 725 over a grid of three threads at a time. Flower 4 is worked in lazy daisy stitch with an extra central straight stitch to each petal, in 725. The centre is French knots in 783. Flower 5 is worked in radiating straight stitches in 210 with a single French knot at the centre in 783. When the embroidery is complete, remove all marking threads, wash gently and press the embroidery. Insert the zip and make up in the usual way.

Embroidered Picture

BY LINDA MCDEVITT

This stunning embroidered picture will take both time and patience, even from the more experienced needle-worker. But when you hang the finished picture on your wall the effect will be well worth the hard work involved.

MEASUREMENTS
Finished size 10 × 14cm (4 × 5½in)

MATERIALS
Single embroidery canvas, 22 threads to 2.5cm (1in), 28 × 32cm (11 × 13in)
Tapestry needle, size 20
Fine tipped indelible marker
1 skein of Anchor stranded cotton in each of the following colours, except Cream (2 steins):

A	Off-white	2	**K**	Mid-blue	850
B	Mid-grey	233	**L**	Mushroom	379
C	Silver Grey	232	**M**	Rust	884
D	Dark Grey	400	**N**	Brown	357
E	Pale Grey	234	**O**	Peach	882
F	Pale Green	875	**P**	Peach Shadow	883
G	Mid-green	877	**Q**	Pink	96
H	Cream	361	**R**	Dark Jade	879
I	Mid-blue	0976	**S**	Turquoise	170
J	Grey-blue	848	**T**	Gold	298

Working the Design
Draw outline of 'photograph' on to the canvas with indelible marker, carefully counting the threads. Make a few guidelines, e.g. skyline and sea. Following the chart and colour key, begin stitching in tent stitch using three strands of stranded cotton. Each square on the chart represents one tent stitch, with the exception of the 'photo corners' (see shaded areas on chart).

TENT STITCH

Stitch the area of the figure, working outwards to cover the background and border.

CORNER STITCH

Finishing
When the work is complete, lay the embroidery face down on several layers of tea towels and press, pulling into shape. If necessary stretch or block to the correct size.

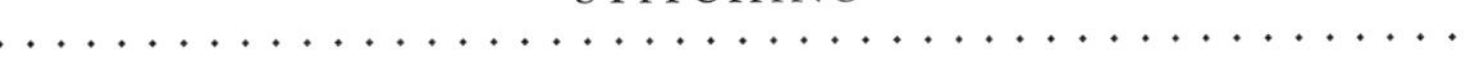

115
110
100
90
80
70
60
50
40
30
20
10
0
10
20
30
40
50
60
70
80
B
A
E
F
G
H
I
C
D
L
O
Q
R
S
T

Christening Sampler

BY ANGELA WAINWRIGHT

This beautiful christening sampler should appeal to all cross-stitch enthusiasts of varying abilities. For those wanting a simpler effect, parts of the design can be omitted.

MATERIALS

Ainring or Hardanger fabric in white, 18 holes to 2.5cm (1in), 33 × 44cm (13 × 17$\frac{1}{2}$in). Tapestry needle, size 26.

DMC stranded embroidery cotton, one skein each of the following shades:

A	Grey 318	E	Green 563
B	Dark Yellow 725	F	Dark Pink 962
C	Medium Yellow 727	G	Pink 605
		H	Brown 407
D	Light Yellow 3078	I	Blue 340

.... Dark Yellow Backstitch

The design is worked using two threads of six-stranded cotton. The bookmark is ready prepared (see page 129 for suppliers) and the fabric has 18 holes to 2.5cm (1in).

The spoon is worked on Hardanger with 22 holes to 2.5cm (1in), using a single thread. (See page 7.)

The pot is worked on a 13cm (5in) square piece of white Ainring or Hardanger with 18 holes to 2.5cm (1in). (See page 7.) See charts on p. 127–131 for full instructions for sampler, spoon and pot. The bookmark is based upon the hearts and church on the sampler (see pp. 128–131). Use at same size as on sampler and space out according to length of the child's and church's name and the relevant date.

METHOD

Names and dates

These are worked in back stitch. Count the number of stitches in the letters you wish to use. Add on one space between each letter and three spaces between each word. Divide this number by half; count along the name to be worked until you reach this number, you now have the centre of the name. Align this with the centre of your work. Begin to work the name with the centre letter you have calculated lining it up with the centre point of the fabric. Work either left or right, whichever you prefer. If you have a spare piece of graph paper it might be a good idea to pencil in the name or dates and place it on the design chart to help you calculate the position.

Other features

'In celebration . . .' worked in blue back stitch using 2 threads.
Child's name worked in brown back stitch using 2 threads.
Church name and dates worked in blue back stitch using 2 threads Outline in dark yellow back stitch all edges of light or medium yellow stitches.
Clock face and window edges in grey back stitch Clock-hands and door in green back stitch.
Internal window lines in medium yellow back stitch.

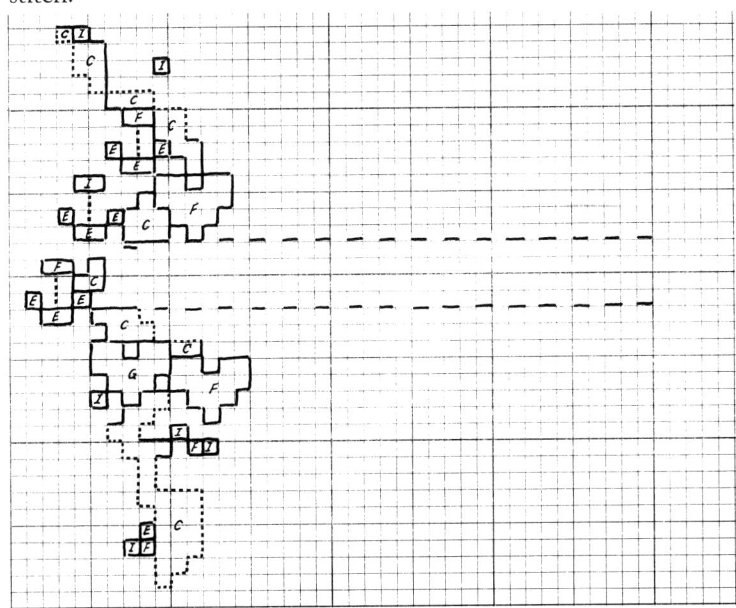

Fig. 1: Pot

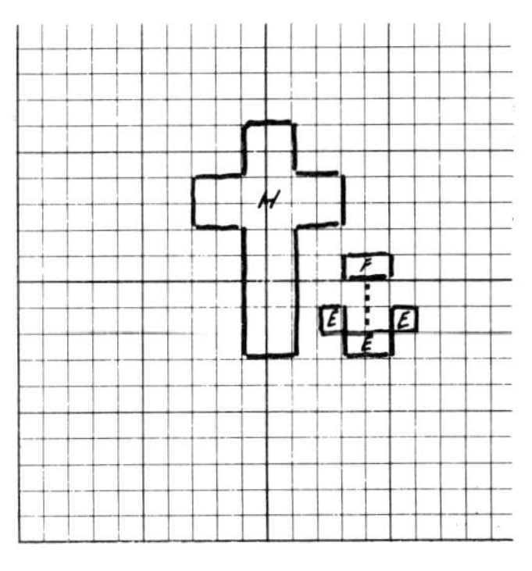

Fig. 2: Spoon

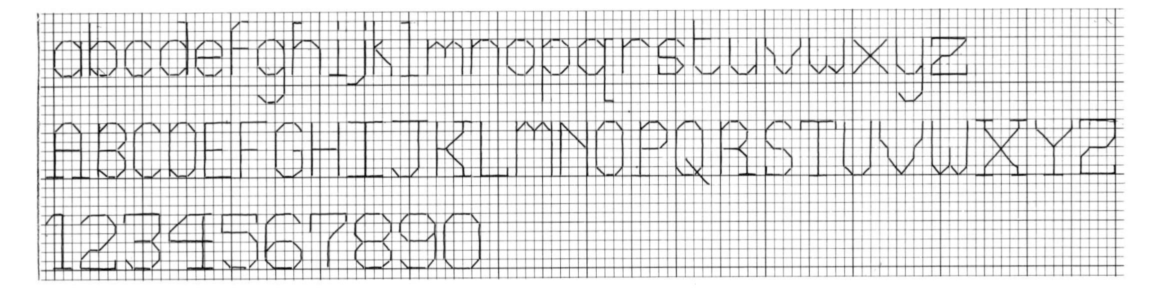

Fig. 3: Numbers and letters guide

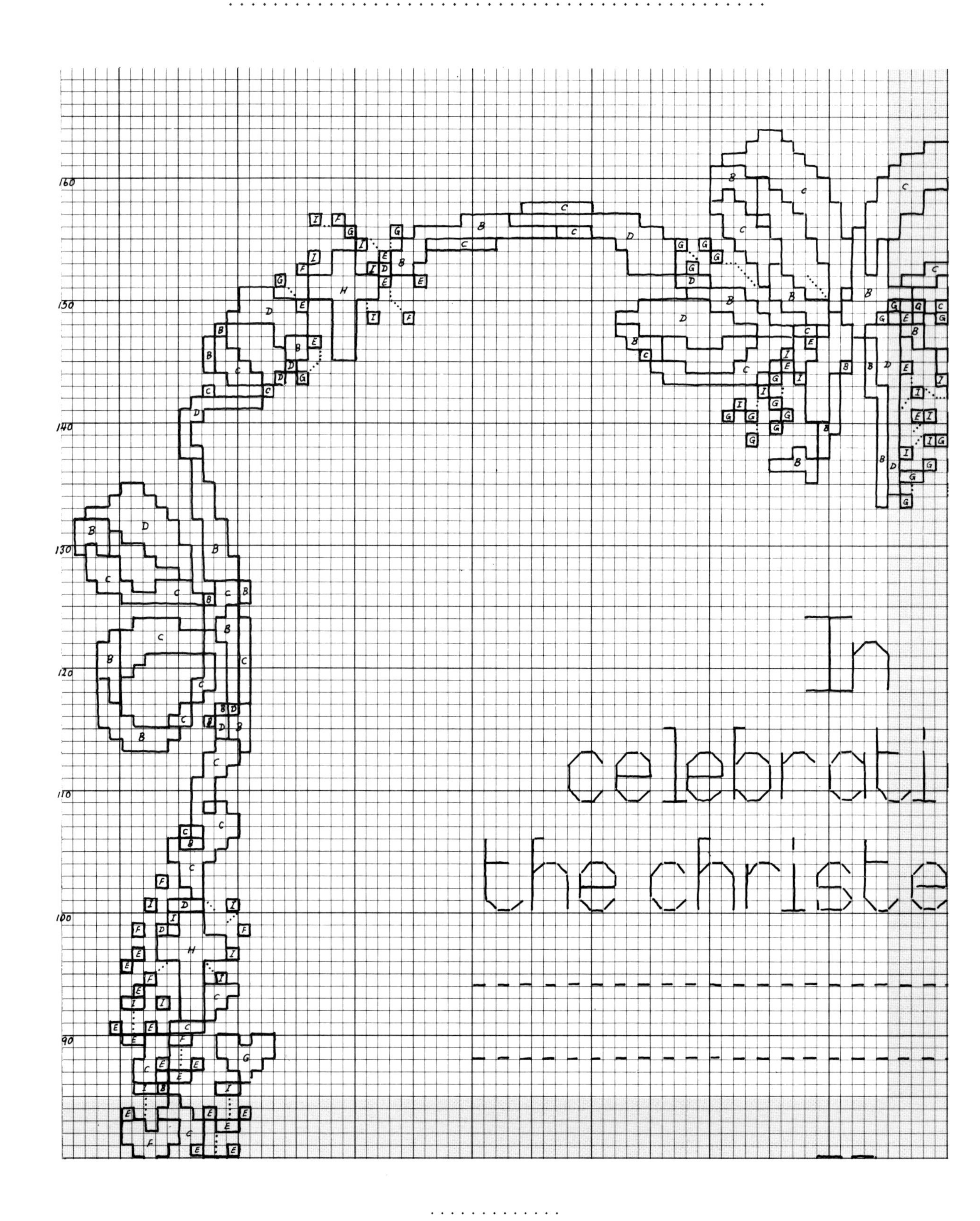
160
150
140
130
120
110
100
90
In
celebrati
the christe

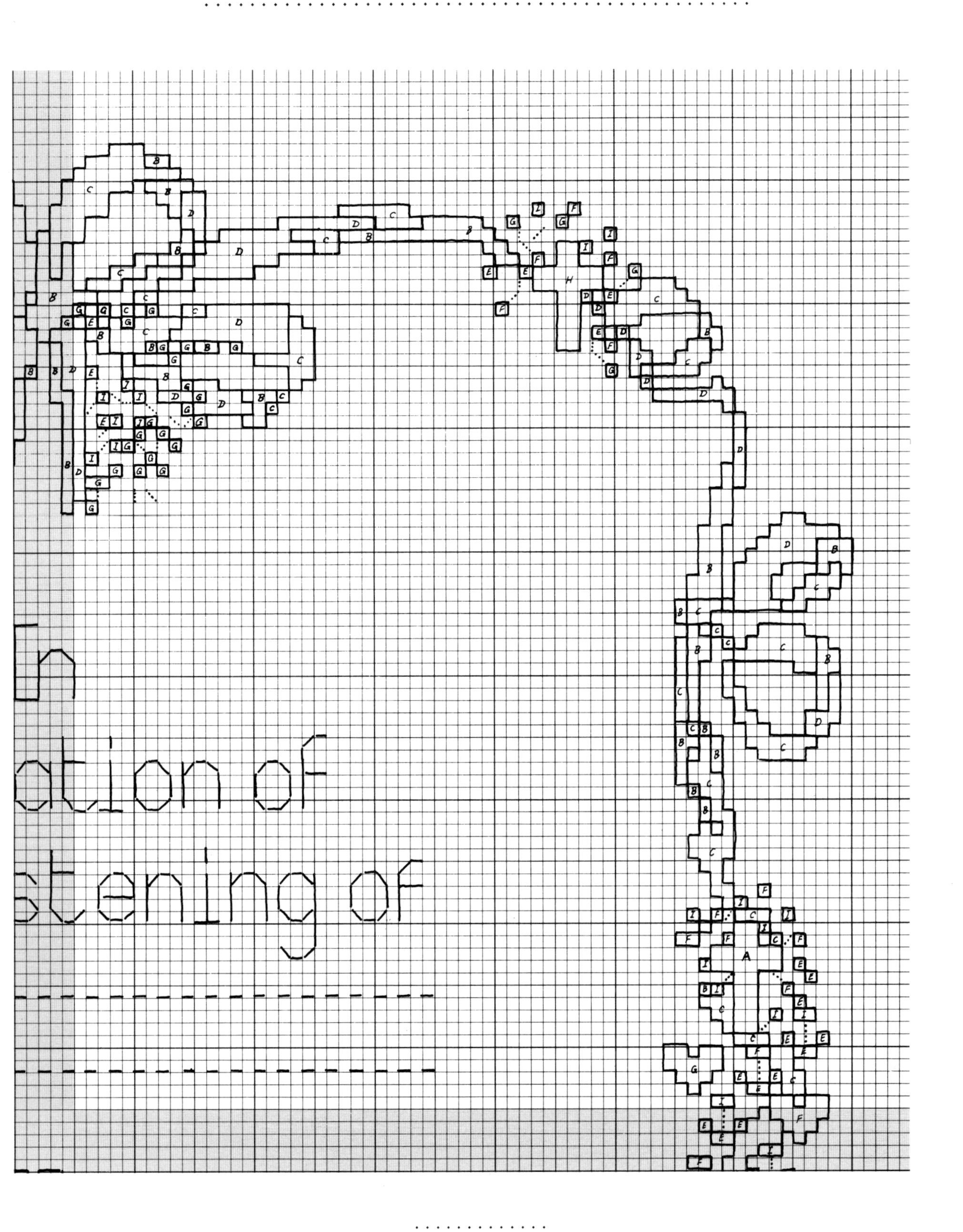
n
ation of
stening of

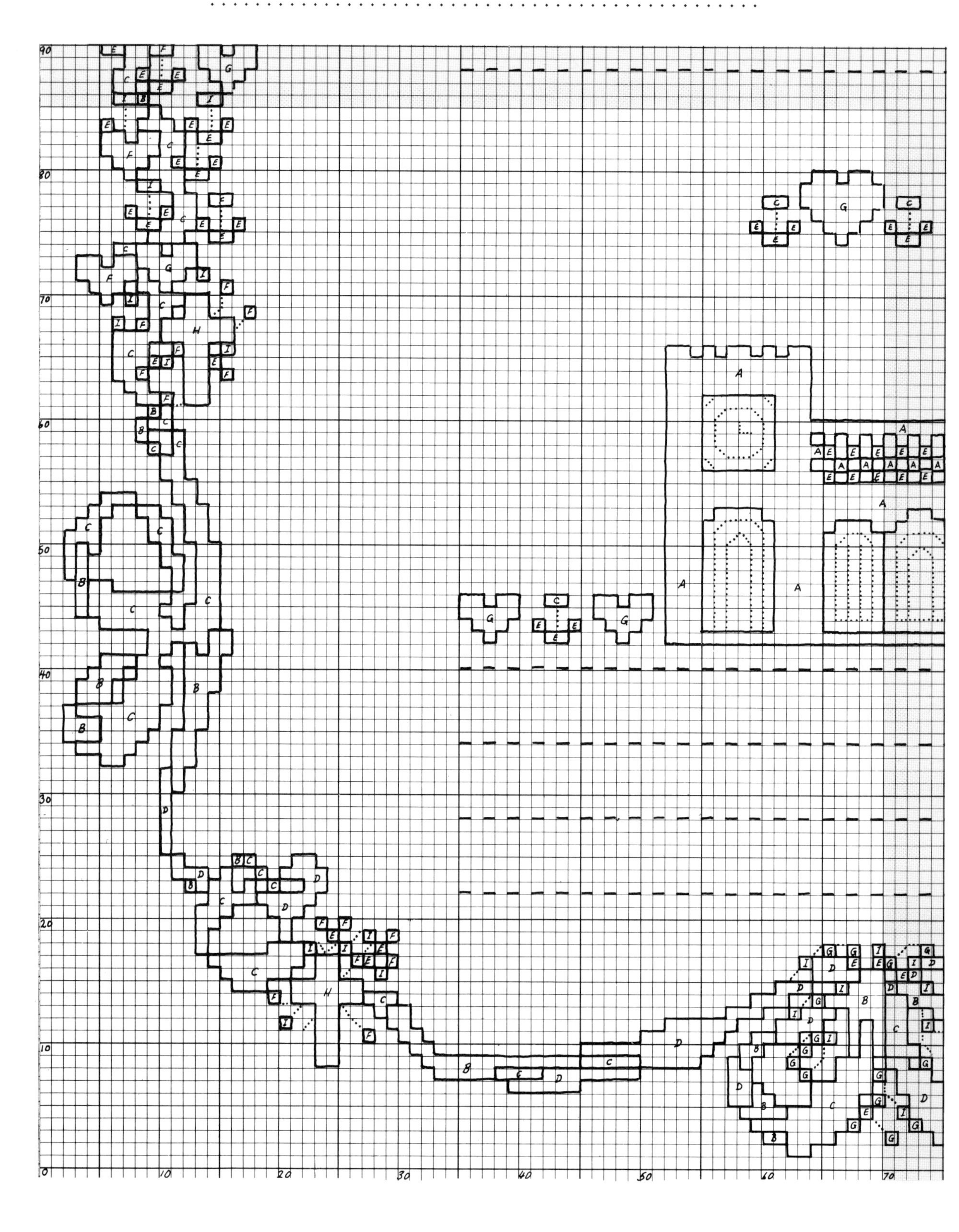

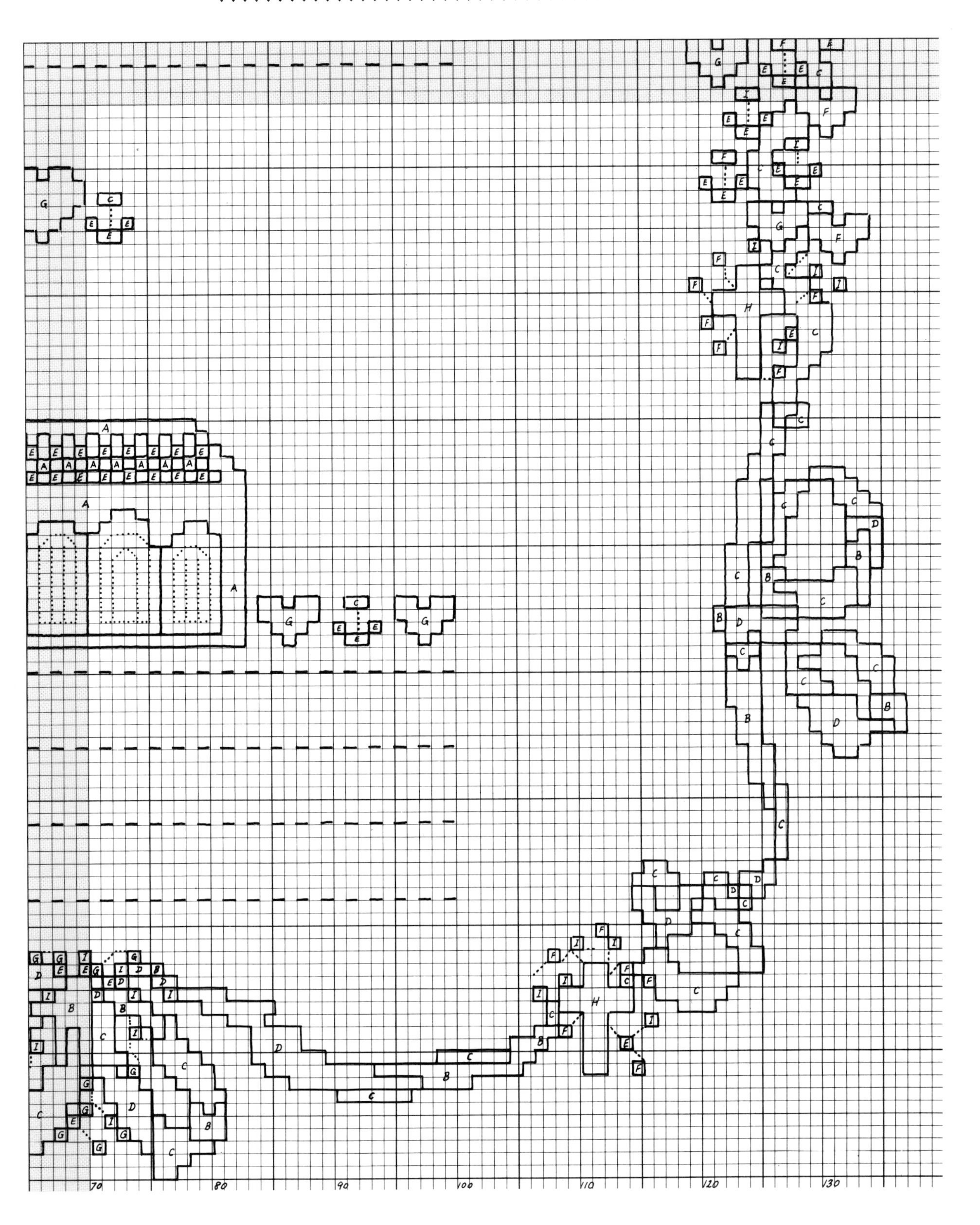

70
80
90
100
110
120
130

The Lion and the Unicorn

BY JULIE HASLER

A visually stunning sampler which achieves its effect by even tension and perfect stitching. This is a project which is probably best suited to the skills of the more experienced stitcher.

You will find cross-stitch a very rewarding and inexpensive hobby. It is very easy to learn, and one of the most versatile and elegant needlecrafts. Following a chart is simple. Each square on the chart represents a square on the fabric. The 'square' on the fabric consists of a block formed by the vertical and horizontal threads. The symbol on the chart represents the colour of the cross-stitch worked over a square of fabric.
1 *Needles* – A small blunt tapestry needle, No. 24 or 26.
2 *Fabric* – Even-weave fabrics, such as Aida, Hardanger and Ainring, on which it is easy to count the threads, are used for cross-stitch. The fabrics are available in a wide choice of colours including white, ecru, black, red, blue, green and yellow. They also come in varying thread counts, the thread count being the number of threads or blocks to each 2.5cm (1in). The fabric to be used for this project is called Jubilee, which is a 28-count fabric. Do not use a fabric which does not have an even weave as this will distort the embroidery either horizontally or vertically.
3 *Threads* – This design has been keyed to shades of DMC six-stranded embroidery cotton. The number of strands you use will depend upon the fabric you decide to work on. Details of this are given in the working.
4 *Embroidery hoop* – 10, 13 or 15cm (4, 5 or 6in) round plastic or wooden hoop with a screw-type tension adjuster is ideal for cross-stitch.
5 *Scissors* – A pair of sharp embroidery scissors is essential, especially if a mistake has to be cut out.

PREPARING TO WORK

To prevent the edges of the fabric unravelling, you can either cover them with a fold of masking tape, or use whip-stitching or machine stitching. Where you make your first stitch is important as it will position the finished design on your fabric. You will need to find the exact centrepoint of the chart by following the arrows on the chart to their intersection. Next, locate the centre of your fabric by folding it in half vertically and then horizontally pinching along the folds. Mark along these lines with basting stitches if you wish. The centre stitch of your design will be where the folds in the fabric meet. It is preferable to begin cross-stitch at the top of the design. To locate the top, count the squares up from the centre of the chart (see pp. 134–135), then count left or right to the first symbol. Next, count the corresponding number of holes up and across from the centre of the fabric and begin at that point. Remember that each square on the chart represents a square on the fabric, and each symbol represents a colour.
To place the fabric in the embroidery hoop, place the area of fabric to be embroidered over the inner ring and gently push the outer ring over it. Gently and evenly pull the fabric, ensuring that it is drum taut in the hoop and the mesh is straight, tightening the screw adjuster as you go. When working, you will find it easier to have the screw in the 'ten-o'clock' position to prevent your thread from becoming tangled in the screw with each stitch. If you are left-handed, have the screw in the 'one-o'clock' position. While working, you will find it necessary to continue to re-tighten the fabric to keep it taut, as tension makes stitching easier, enabling the needle to be pushed through the holes without piercing the

fibres of the fabric. When working with stranded cotton, always separate the strands and place them together again before threading your needle and beginning to stitch. Never double the thread. For example if you need to use two strands, use two separate strands, not one doubled up. These simple steps will allow for much better coverage of the fabric.

TECHNIQUES

Cross-stitch

To begin the stitch, bring the needle up from the wrong side, through the hole in the fabric (as in fig 1) at the left end of a row of stitches of the same colour. Fasten the thread by holding a short length of thread on the underside of the fabric, securing it with the first two or three stitches made, as in fig 2. Never use knots to fasten the thread as this will create a bumpy back surface and prevent your work from laying flat when completed. Next, bring the needle across one square (or block) to the right and one square above on a left to right diagonal and insert the needle as in fig 1. Half the stitch is now completed. Continue in this way until the end of the row is reached. Your stitches should be diagonal on the right side of the fabric and vertical on the wrong side. Complete the upper half of the stitch by crossing back from right to left to form an 'x' as in fig 3. Work all the stitches in the row by completing all the x's, as in fig 4. Cross-stitch can also be worked by crossing each stitch as you come to it, as you would do for isolated stitches. This method works just as well as the previous one – it is really a personal preference. Work vertical rows of stitches as shown in fig 5.

Finish all threads by running your needle under four or more stitches on the wrong side of the work as shown in fig 6 and cut close.

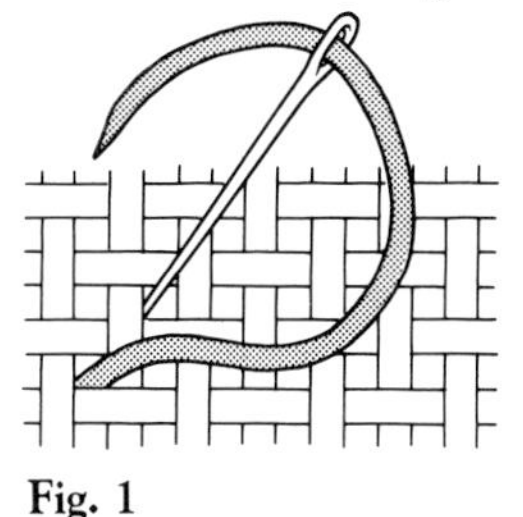

Fig. 1

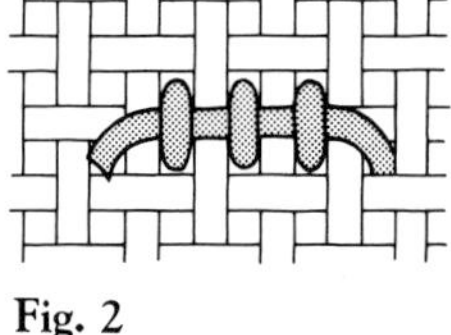

Fig. 2

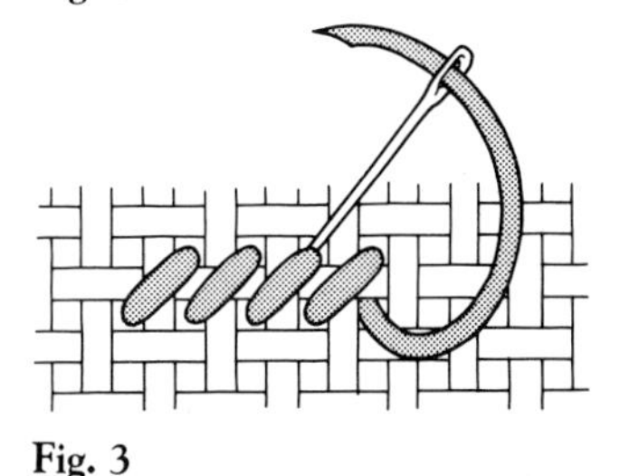

Fig. 3

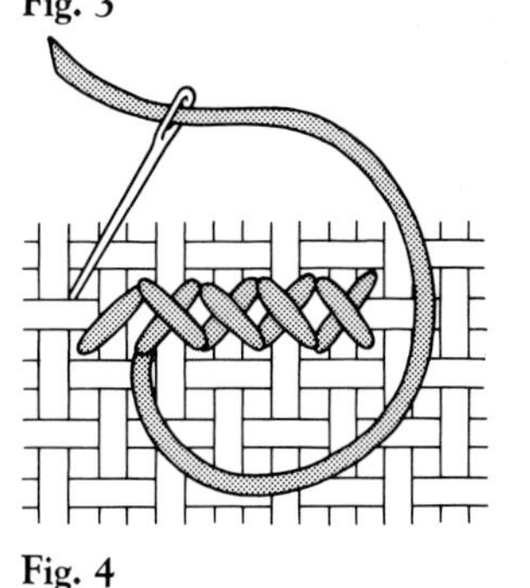

Fig. 4

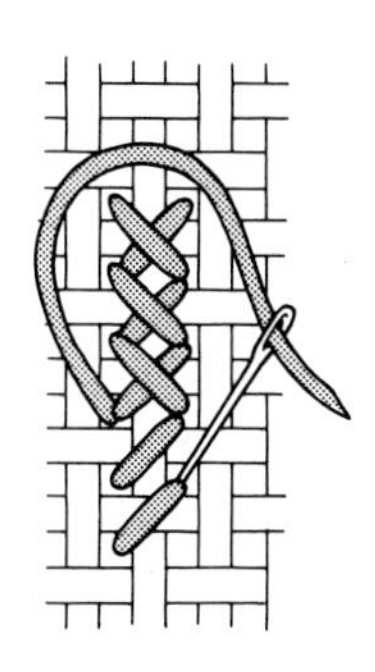

Fig. 5

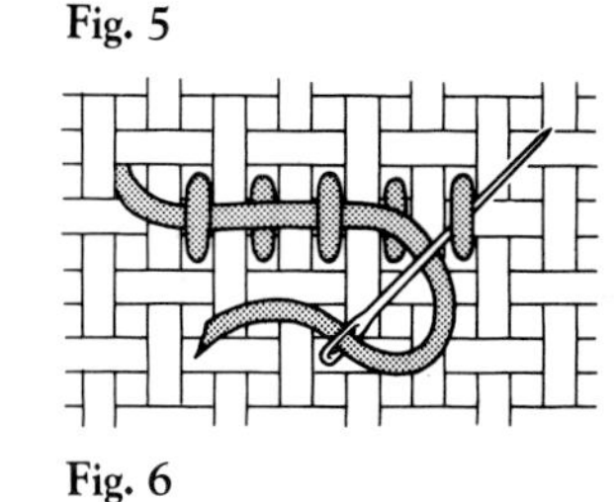

Fig. 6

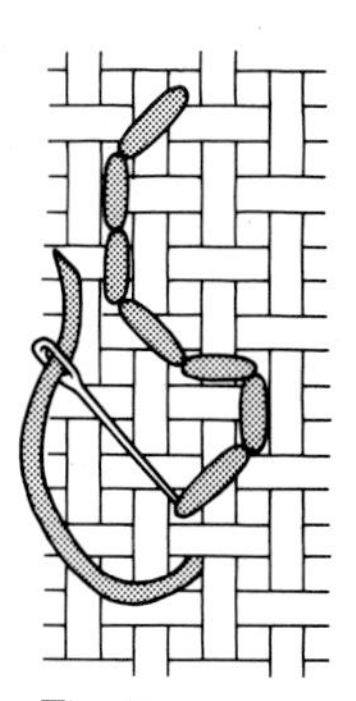

Fig. 7

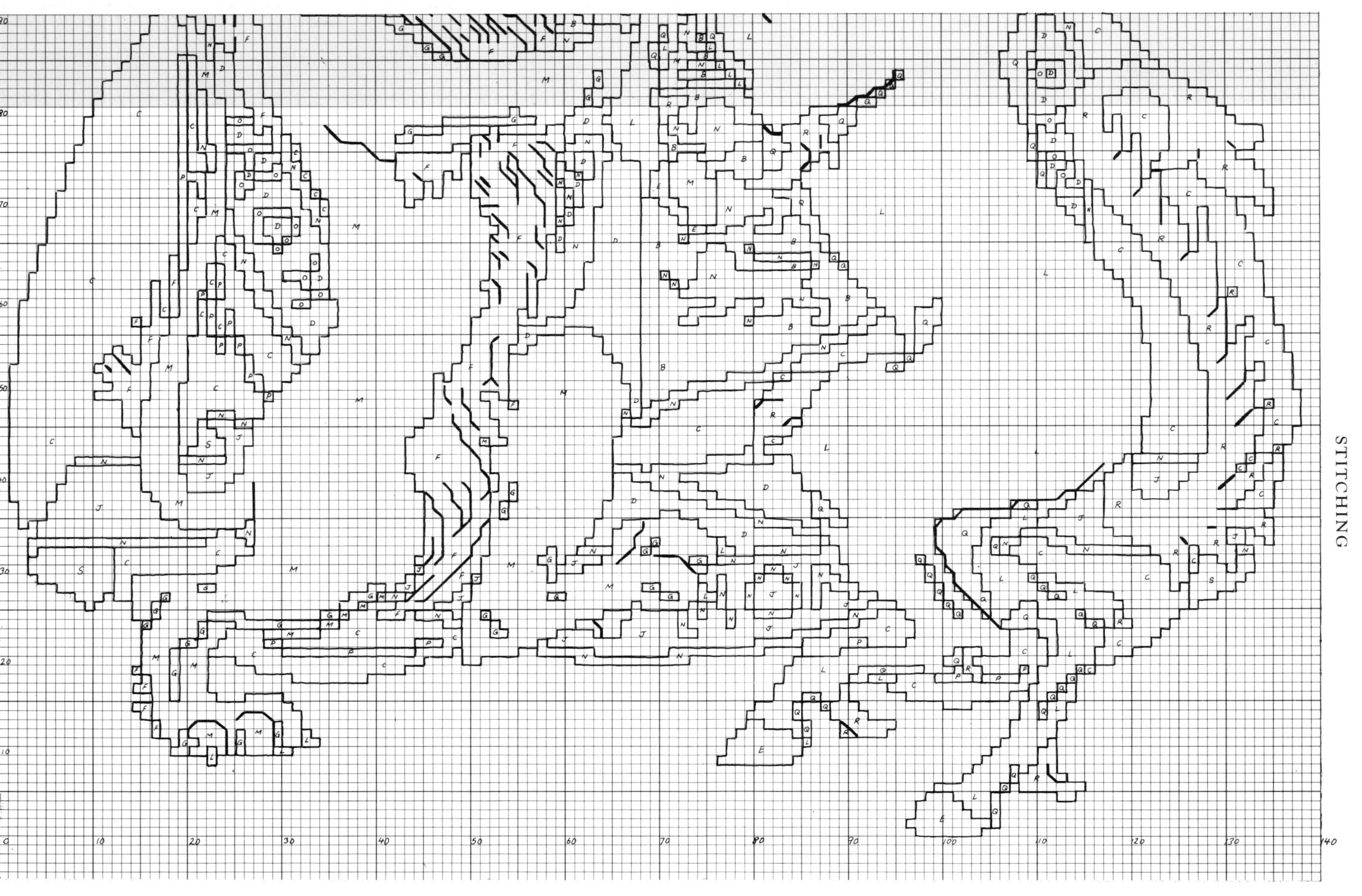

A – 782 Gold Thread (backstitch) **B** – 304 Medium Christmas Red **C** – 210 Medium Lavender **D** – 809 Delft Blue **E** – 453 Light Shell Grey (backstitch 317)
F – 782 Medium Topaz Brown (backstitch 801) **G** – 729 Medium Old Gold **H** – 725 Topaz Yellow **I** – 310 Black **J** – 818 Baby Pink **K** – 353 Peach **L** – White
M – 677 Very Light Old Gold (backstitch 729) **N** – 726 Light Topaz Yellow **O** – 3348 Light Yellow Green **P** – 208 Very Dark Lavender **Q** – 415 Pale Grey
(backstitch 317) **R** – 746 Off White (backstitch 437) **S** – 899 Medium Pink **T** – 746 Off White

BASIC BACKSTITCH

Basic backstitch is used to outline features and finer details in some cross-stitch designs, to emphasize the colours and make them stand out. Work any backstitch when your cross-stitch embroidery has been completed. Use one strand less than that used in the embroidery. For example, if three strands of stranded cotton have been used to work the cross-stitch embroidery, use two strands for the backstitching. If only one strand of stranded cotton is used to work the cross-stitch embroidery, one strand is also used for the backstitching. Backstitch is worked from hole to hole and can be stitched in diagonal, vertical or horizontal lines as shown in fig 7. Always take care not to pull the stitches too tight, otherwise the contrast of colour will be lost against the cross-stitches. Finish off the threads as for cross-stitch.

Where backstitch is required, it is indicated on the chart by the use of a bolder outline.

USEFUL TIPS

1 When you are stitching, it is important not to pull the fabric out of shape. You can accomplish this by working the stitches in two motions, straight up through a hole in the fabric and then straight down ensuring the fabric remains taut. Make sure that you don't pull the thread tight – it should be snug, but not tight. If you use this method, you will find the thread will lie just where you want it to and not pull your fabric out of shape.

2 If the thread becomes twisted while working, drop the needle and let it hang down freely. It will then untwist itself. Do not continue working with twisted thread as it will appear thinner and not cover the fabric as evenly.

3 Never leave the needle in the design area of your work when not in use. No matter how good the needle might be, it could rust in time, and may mark your work permanently.

4 Do not carry thread across an open expanse of fabric. If you are working separate areas of the same colour, finish off and begin again. Loose threads, especially dark colours, will be visible from the right side of your work when the project is completed.

5 When you have completed your cross-stitch embroidery, it will need to be pressed. To protect your work, place the embroidery right side down on to a soft towel and cover the reverse side with a thin, slightly damp cloth before pressing.

This beautiful design is based upon the nursery rhyme *The Lion and the Unicorn*.

MATERIALS

Jubilee fabric with 28 stitches to 2.5cm (1in) in a colour of your choice, 51 × 61cm (20 × 24in)
Piece of mounting board approximately 41 × 51cm (16in × 20in)
Masking tape
DMC stranded cotton in the colours indicated

Directions

1 Complete the cross-stitch embroidery centrally on the Jubilee fabric, sewing over two blocks of the fabric, therefore effectively making it a 14-count embroidery, using two strands of the stranded cotton for the cross-stitch, and one for the backstitch. When completed, press flat.

2 To mount your embroidery ready for framing, place it face down on a clean, flat surface and position the mounting board centrally on top. Rest a heavy book or similar weight on top to prevent it moving out of place.

3 Mitre the corners very carefully, cutting the fabric a fraction away from the corners of the board. Fold one edge of the fabric over the mounting board, ensuring that the fabric is absolutely straight, and secure it with pins along the edge of the board. Fix the opposite edge in the same way, ensuring the fabric is taut and straight on the board. Secure the edges of the fabric to the back of the mounting board using the masking tape, then remove the pins. Repeat this procedure for the remaining two edges.

4 The embroidered picture is now ready to be framed. Unless you are experienced in this craft, it is best to take your work to a professional framer. If you wish to have glass in the frame, the non-reflective glass gives a far superior effect.